Otto T. Bruhns

Advanced Mechanics of Solids

Springer
Berlin
Heidelberg
New York
Barcelona
Hongkong
London
Mailand
Paris
Tokio

Engineering ONLINE LIBRARY

http://www.springer.de/engine-de/

Otto T. Bruhns

Advanced
Mechanics of Solids

With 81 Figures

Springer

Professor Dr.-Ing. Otto T. Bruhns
Ruhr-Universität Bochum
Institut für Mechanik
Universitätsstr. 150
D-44780 Bochum
e-mail: *bruhns@tm.bi.ruhr-uni-bochum.de*

ISBN 978-3-642-07850-7

Cataloging-in-Publication Data applied for

Bibliographic information published by Die Deutsche Bibliothek
Die Deutsche Bibliothek lists this publication in the Deutsche Nationalbibliografie;
detailed bibliographic data is available in the
Internet at <http://dnb.ddb.de>

Springer-Verlag Berlin Heidelberg New York
a member of BertelsmannSpringer Science + Business Media GmbH

http://www.springer.de

© Springer-Verlag Berlin Heidelberg 2010
Printed in Germany

Cover-Design: deblik, Berlin
Printed on acid-free paper

Preface

Mechanics, and in particular, the mechanics of solids, forms the basis of all engineering sciences. It provides the essential foundations for understanding the action of forces on bodies, and the effects of these forces on the straining of the body on the one hand, and on the deformation and motion of the body on the other. Thus, it provides the solutions of many problems with which the would-be engineer is going to be confronted with on a daily basis.

In addition, in engineering studies, mechanics has a more vital importance, which many students appreciate only much later. Because of its clear, and analytical setup, it aids the student to a great extent in acquiring the necessary degree of abstraction ability, and logical thinking, skills without which no engineer in the practice today would succeed.

Many graduates have confirmed to me that learning mechanics is generally perceived as difficult. On the other hand, they always also declared that the preoccupation with mechanics made an essential contribution to their successful education. Besides, as far as my experience goes, this success does not depend very much on the inclusion of special chapters, or the knowledge of particular formulae. Rather, it is important that to a sufficient degree, one has learned how to logically describe a given physical phenomenon, starting from the preconditions. And that from this description one can derive rules for related phenomena, and also rules for lay out design, for dimensioning, etc. similarly supported structures.

This book follows closely the didactic line of the course Mechanics of Solids, which I have taught since the introduction of the master's program "Computational Engineering" at the Ruhr-University Bochum. The story of this book, however, goes back a long way. First sketches and a first version of the book go back to my time as a young professor at the university of Kassel, which already more than twenty years ago was the first and only German university to introduce consecutive engineering programs. Being the professor responsible for this program, I thought about the necessary contents. Many of my ideas from that time have been used in this book.

An essential starting point for designing the course was the fact that many students taking a master's program already have, to a certain degree, background knowledge in mechanics. This also applies to this book. It is assumed that the reader is familiar with the elementary methods of statics, in particular the statics of statically determinate structures and systems. In addition, elementary knowledge on strength of materials is assumed. Thus, it is supposed that the terms stress, and strain as well as deformation, and displacement are well-known and don't need to be explained again. Furthermore, the reader should also be familiar with the relations for calculating the normal stress for the simple bending of prismatic beams.

Starting from this basic knowledge, the principles of linear continuum mechanics on which much of the material in the book depends are presented in Chapter 1. Complementary to that is Chapter 2 which deals with the constitutive equations for linear elastic material behaviour. A brief overview of classical strength criteria provides the relevant limits of elastic calculation methods for the practice. Essential in the age of computers, and of discretizing methods such as the Finite Element Method for calculating structures, is the treatment given the fundamental principles underlying these methods.

In Chapter 3, these general relations are specialized to one-dimensional structures (beams). Starting from well-known relations, much emphasis is placed on the description of the state of stress in thin-walled cross sections. Alongside, topics such as the influence of distributed loads or the distribution of stresses in non-prismatic structures are studied.

The torsion of prismatic beams is presented in Chapter 4. In addition to the description of solid cross sections, particular attention is paid here again to thin-walled cross sections. Included in these studies is the description of torsion with restrained warping.

Chapter 5 is concerned with the description of the state of stress and the deformations of curved beams with small and large curvature.

Energy methods introduced at the beginning are applied to beam-like structures in Chapter 6. Using the reciprocity theorems of Betti and Maxwell as well as the theorems of Castigliano and Engesser, various methods for solving statically indeterminate systems are presented.

After the one-dimensional structures, two-dimensional structures are studied in Chapters 7 and 8. First, in Chapter 7, the governing equations as well as the classical solutions for both special cases plane stress and plane strain are addressed. Chapter 8 then deals with disks, plates, and shells, where attention is restricted to the membrane theory of shells of revolution.

This part of the book concludes with Chapter 9 on the buckling and stability of beams. As extension of the well-known classical solutions of the Euler columns, energy methods and approximate solutions based on these methods are given special emphasis.

These first chapters are supplemented by chapters 10 to 12, where in addition, and unlike many other text books on solid mechanics, the essential and fundamental calculation methods for dealing with systems with a finite degree of freedom are presented.

Essentially, there were two reasons that prompted me to include these chapters from dynamics in this course. On one hand, the engineer in practice is confronted in increasing degree with problems dealing with the dynamic behaviour of structures. On the other hand, most of the methods from statics are also easily applicable to dynamics. This is the case e.g. for the energy methods treated in Chapter 6, and in particular, for the influence coefficients introduced there, and for many approximation methods that can be deduced from these energy methods.

The book is structured such that in each chapter the theoretical considerations are accompanied by several illustrative examples demonstrating the application of these results. At the end of each chapter, I have included additional problems for the reader. The solutions of these problems are given in Chapter 13.

In preparing the manuscript, I had valuable assistance from Ndzi C. Bongmba, who read the original manuscript and suggested several improvements of the presentation, in particular, of the examples and problems. I gratefully acknowledge his contribution.

Bochum, July 2002   *Otto Bruhns*

Contents

1. Basic Concepts of Continuum Mechanics

1.1 General Remarks

In Chapter 1, we present general concepts and definitions that are fundamental to many of the topics discussed in this course.

Mechanics has been defined as the study of forces and motions. It is easy to define motions as the change in position of a body, in time, and with respect to some frame of reference. The total force $\mathbf{F}$ on a body $\mathcal{B}$ is the vector sum of all the forces exerted on it by different actions.

The purpose of the majority of structures or structural members designed by engineers is to transmit these forces; and it is of great importance for the designer to know the manner in which the force is transmitted in each member, because it may be that the mode of internal distribution of the force - conditioned by the shape and dimensions of the member - is such that failure of the material of which the member is made may occur at some point or points in the member. It is necessary, therefore, to consider how a force is transmitted through an element of material.

1.2 Stresses

Consider a solid body under the action of a system of forces in equilibrium, e.g. distributed forces $\mathbf{f}$, and concentrated forces $\mathbf{F}_i$, $i = 1, 2, \ldots, n$, respectively, acting on the outer surface of this body, and concentrated reactions acting at the two supports as sketched in Fig. 1.1. When we cut this body into two pieces, passing an arbitrary surface through it, each of the two pieces will be in equilibrium if an additional force is applied to it. In the uncut body this force must be transmitted through the imaginary surface, and it is plausible to assume that each area element ΔA makes its contribution $\Delta \mathbf{F}$. The limit of the quotient $\Delta \mathbf{F} / \Delta A$ for vanishing ΔA is called the stress

$$\mathbf{s}(\mathbf{n}) = \lim_{\Delta A \to 0} \frac{\Delta \mathbf{F}}{\Delta A} = \frac{d\mathbf{F}}{dA} \tag{1.1}$$

in the element, where the orientation of area dA is designated by the unit vector $\mathbf{n}$. $\mathbf{s}(\mathbf{n})$ is a vector and may be resolved, in general, into components parallel to $\mathbf{n}$ and perpendicular to $\mathbf{n}$. These are the projected stresses, namely, normal stress σ_{nn}

$$\sigma_{nn} = \sigma(\mathbf{n})\mathbf{n} = (\mathbf{s} \cdot \mathbf{n})\mathbf{n}, \tag{1.2}$$

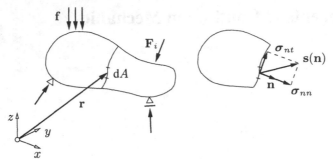

Fig. 1.1 Section surface and related stress vector s

and shear stress σ_{nt}

$$\sigma_{nt} = \mathbf{s} - \sigma(\mathbf{n})\mathbf{n} = \mathbf{n} \times (\mathbf{s} \times \mathbf{n}),\tag{1.3}$$

with $\sigma_{nt} \cdot \mathbf{n} = 0$. Obviously, σ_{nt} can be once more resolved into two components in two arbitrary directions in the area element dA.

When different sections are passed through the same point characterized by vector $\mathbf{r}$, different stresses are found in areas dA of different orientation. Especially, in a system of rectangular cartesian coordinates, when $\mathbf{n} = \mathbf{e}_x$,

$$\mathbf{s}_x = \sigma_{xx}\mathbf{e}_x + \sigma_{xy}\mathbf{e}_y + \sigma_{xz}\mathbf{e}_z = \sigma_{xx} + \sigma_{xt}.\tag{1.4}$$

Using Einstein's summation convention, Eq. (1.4) may be written in a much shorter form

$$\mathbf{s}_x = \sigma_{xi}\mathbf{e}_i, \quad i = x, y, z.\tag{1.5}$$

In an analogous manner, we may consider two additional sections perpendicular to the y and z axes

$$\mathbf{s}_y = \sigma_{yi}\mathbf{e}_i,$$
$$\mathbf{s}_z = \sigma_{zi}\mathbf{e}_i.\tag{1.6}$$

It can be shown that in an Eulerian space the state of stress at some point is uniquely determined by these three stress vectors from three orthogonal sections, or their 9 components, which are the 9 components σ_{ik}, $i, k = x, y, z$ of the stress tensor σ. The four sections described above constitute a small tetrahedron at point $\mathbf{r}$ with three faces lying in planes perpendicular to the x, y and z axes, and a fourth one with area dA characterized by unit vector $\mathbf{n}$. Thus the three vector equations (1.5) and (1.6) may be summarized to give

$$\mathbf{s}(\mathbf{n}) = \mathbf{s}_i n_i,\tag{1.7}$$

and further

$$\boxed{\mathbf{s} = \sigma^T\mathbf{n}, \quad s_k = \sigma_{ik}n_i,}\tag{1.8}$$

where now the second-rank stress tensor σ is defined as a linear mapping, and

$$\sigma = \sigma_{ik}\mathbf{e}_i \otimes \mathbf{e}_k, \quad i, k = x, y, z,\tag{1.9}$$

with the symbol $\otimes$ designating the tensor product (dyadic product) of two vectors. It is sometimes usual to write this in the form of a matrix

$$\sigma = \begin{pmatrix} \sigma_{xx} & \sigma_{xy} & \sigma_{xz} \\ \sigma_{yx} & \sigma_{yy} & \sigma_{yz} \\ \sigma_{zx} & \sigma_{zy} & \sigma_{zz} \end{pmatrix}. \tag{1.10}$$

We note, however, that the σ_{ik} are the coordinates (components) of the stress tensor σ. Herein, the first subscript indicates the section in which the stress acts, and the second subscript indicates the direction of the stress.

As it is usual in classical (continuum) mechanics, we exclude distributed couples acting on a material particle (Boltzmann's theorem). With this axiom, the symmetry of the stress tensor

$$\sigma = \sigma^{\mathrm{T}} \quad \rightarrow \quad \sigma_{ik} = \sigma_{ki} \tag{1.11}$$

follows from the equilibrium of the moments of the forces acting on a volume element $dx\,dy\,dz$. Equation (1.11) reduces the number of independent components from 9 to 6.

First, we consider the more specific case of plane or biaxial stresses, where the stress components in the z direction are assumed to vanish

$$\sigma_{zx} = \sigma_{zy} = \sigma_{zz} = 0. \tag{1.12}$$

For this plane stress case, we will now discuss the influence of the direction of the section line on these stresses, i.e. we will discuss the transformation of these stresses under the influence of a rotation of our coordinate system about the z axis.

Consider a small triangular prism as shown in Fig. 1.2.

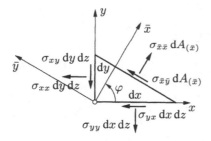

Fig. 1.2
Transformation of the stresses

Let φ be the angle of rotation, and dA be the area of the inclined face. Then, the areas of the other faces are $dA\cos\varphi$ and $dA\sin\varphi$, and the equilibrium of forces in direction $\bar{x}$ and $\bar{y}$ yields the first two of the following equations

$$\sigma_{\bar{x}\bar{x}} = \sigma_{xx}\cos^2\varphi + \sigma_{yy}\sin^2\varphi + 2\sigma_{xy}\sin\varphi\cos\varphi$$

$$\sigma_{\bar{x}\bar{y}} = (\sigma_{yy} - \sigma_{xx})\sin\varphi\cos\varphi + \sigma_{xy}(\cos^2\varphi - \sin^2\varphi) \tag{1.13}$$

$$\sigma_{\bar{y}\bar{y}} = \sigma_{xx}\sin^2\varphi + \sigma_{yy}\cos^2\varphi - 2\sigma_{xy}\sin\varphi\cos\varphi.$$

Equation $(1.13)_3$ is derived from a prism containing a face $\bar{y} = \text{const.}$ These equations may be written in a more compact form

$$\sigma_{\bar{x}\bar{x}} = \frac{1}{2}\left(\sigma_{xx} + \sigma_{yy}\right) + \frac{1}{2}\left(\sigma_{xx} - \sigma_{yy}\right)\cos 2\varphi + \sigma_{xy}\sin 2\varphi$$
$$\sigma_{\bar{y}\bar{y}} = \frac{1}{2}\left(\sigma_{xx} + \sigma_{yy}\right) - \frac{1}{2}\left(\sigma_{xx} - \sigma_{yy}\right)\cos 2\varphi - \sigma_{xy}\sin 2\varphi \qquad (1.14)$$
$$\sigma_{\bar{x}\bar{y}} = \sigma_{\bar{y}\bar{x}} = \qquad -\frac{1}{2}\left(\sigma_{xx} - \sigma_{yy}\right)\sin 2\varphi + \sigma_{xy}\cos 2\varphi \,.$$

From these transformation formulas, we can conclude the following relations

1. A rotation of $\varphi = n\pi$ does not alter the components.
2. A rotation of $\varphi = \pi/2$ causes

$$\sigma_{\bar{x}\bar{x}} = \sigma_{yy}, \quad \sigma_{\bar{y}\bar{y}} = \sigma_{xx}, \quad \sigma_{\bar{x}\bar{y}} = -\sigma_{xy}. \qquad (1.15)$$

3. Extreme values of $\sigma_{\bar{x}\bar{x}}$ and $\sigma_{\bar{y}\bar{y}}$, and vanishing shear stresses $\sigma_{\bar{x}\bar{y}}$ are attained for an inclination of

$$\varphi_0 = \frac{1}{2}\arctan\frac{2\sigma_{xy}}{\sigma_{xx} - \sigma_{yy}}. \qquad (1.16)$$

The orthonormal base vectors (e_1, e_2) related with these directions form the principal axes of the stress tensor with the extreme values

$$\sigma_{1,2} = \frac{1}{2}\left(\sigma_{xx} + \sigma_{yy}\right) \pm \frac{1}{2}\sqrt{\left(\sigma_{xx} - \sigma_{yy}\right)^2 + 4\sigma_{xy}^2}, \qquad (1.17)$$

where $\sigma_1 \geq \sigma_2$. Since the solution of Eq. (1.16) is in general not unique, the following table may help in finding the magnitudes of the inclination angle

$\sigma_{xx} \geq \sigma_{yy}$	$\sigma_{xy} \geq 0$	$0 \leq \varphi \leq \dfrac{\pi}{4}$
	$\sigma_{xy} \leq 0$	$-\dfrac{\pi}{4} \leq \varphi \leq 0$
$\sigma_{xx} \leq \sigma_{yy}$	$\sigma_{xy} \geq 0$	$\dfrac{\pi}{4} \leq \varphi \leq \dfrac{\pi}{2}$
	$\sigma_{xy} \leq 0$	$\dfrac{\pi}{2} \leq \varphi \leq \dfrac{3}{4}\pi$

Table 1.1 Orientation of principal axes

4. Analogously, we find

$$\varphi_1 = \frac{1}{2}\arctan\frac{\sigma_{yy} - \sigma_{xx}}{2\sigma_{xy}} \qquad (1.18)$$

for the angle of inclination of the section with extreme shear stresses

$$|\sigma_{\bar{x}\bar{y}}|_{\max} = \frac{1}{2}\sqrt{\left(\sigma_{xx} - \sigma_{yy}\right)^2 + 4\sigma_{xy}^2} = \frac{1}{2}\left(\sigma_1 - \sigma_2\right). \qquad (1.19)$$

We realize that since

$$\varphi_1 = \varphi_0 \pm \frac{\pi}{4}, \qquad (1.20)$$

the direction of maximum shear stresses is inclined by $45°$ to the principal axes. Moreover,

$$\sigma_{\bar{x}\bar{x}} = \sigma_{\bar{y}\bar{y}}|_{\varphi=\varphi_1} = \frac{1}{2}\left(\sigma_{xx} + \sigma_{yy}\right) = \frac{1}{2}\left(\sigma_1 + \sigma_2\right). \tag{1.21}$$

5. If the principal values of the stresses σ_1, σ_2 are given, we determine the stresses in sections inclined by an angle α from

$$\left.\begin{array}{c}\sigma_{xx}\\ \sigma_{yy}\end{array}\right\} = \frac{1}{2}\left(\sigma_1 + \sigma_2\right) \pm \frac{1}{2}\left(\sigma_1 - \sigma_2\right)\cos 2\alpha$$

$$\sigma_{xy} = -\frac{1}{2}\left(\sigma_1 - \sigma_2\right)\sin 2\alpha. \tag{1.22}$$

6. We further realize that the quantities

$$\bar{S}_1 = \sigma_{xx} + \sigma_{yy} = \sigma_1 + \sigma_2$$
$$\bar{S}_2 = \sigma_{xx}^2 + 2\sigma_{xy}^2 + \sigma_{yy}^2 = \sigma_1^2 + \sigma_2^2 \tag{1.23}$$
$$\bar{S}_3 = \sigma_{xx}^3 + 3\sigma_{xy}^2(\sigma_{xx} + \sigma_{yy}) + \sigma_{yy}^3 = \sigma_1^3 + \sigma_2^3$$

are independent from any rotation and thus invariants of the plane stress state.

Now, the foregoing considerations will be transferred to general (three-dimensional) states of stress. The question therefore arises, whether it is possible to describe any given state of stress as a superposition of three normal stresses on mutually perpendicular planes, such that all three shear stresses are zero simultaneously, i.e. that the stress tensor is described by

$$\sigma = \begin{pmatrix} \sigma_1 & 0 & 0 \\ 0 & \sigma_2 & 0 \\ 0 & 0 & \sigma_3 \end{pmatrix}. \tag{1.24}$$

Let n be a unit vector parallel to such an axis; then, from Eqs. (1.3) and (1.8), we find

$$s - \sigma(n)n = \sigma n - \sigma(n)n = 0, \tag{1.25}$$

or, equivalently,

$$(\sigma - \sigma(n)1)n = 0. \tag{1.26}$$

This relation represents a system of three linear homogeneous equations for the components of the unit vector n.

In order that $n \neq 0$, it is necessary that

$$\det(\sigma - \sigma 1) = 0 \quad \rightarrow \quad \det(\sigma_{ik} - \sigma\delta_{ik}) = 0 \tag{1.27}$$

holds, where 1 is the unit tensor, and

$$\delta_{ik} = \begin{cases} 1 & \text{for } i = k \\ 0 & \text{for } i \neq k \end{cases} \tag{1.28}$$

is called the Kronecker Delta.

This cubic equation (Eq. 1.27)

$$\boxed{\sigma^3 - S_1\sigma^2 - S_2^*\sigma - S_3^* = 0} \qquad (1.29)$$

has three roots, which are the values of σ for which Eq. (1.24) holds. Such roots are known in general as the eigenvalues of the matrix σ, and in this particular case as the principal stresses. To each σ_i, $i = 1, 2, 3$, there corresponds an eigenvector $\mathbf{n}^{(i)}$; an axis directed along an eigenvector is called a principal axis of stress.

Here, S_1, S_2^*, and S_3^* are the so-called principal invariants of the tensor σ

$$
\begin{aligned}
S_1 &= \sigma_{ii} = \sigma_1 + \sigma_2 + \sigma_3 \\
S_2^* &= \tfrac{1}{2}(\sigma_{ik}\sigma_{ki} - \sigma_{ii}^2) = -(\sigma_1\sigma_2 + \sigma_2\sigma_3 + \sigma_3\sigma_1) \\
S_3^* &= \det\sigma_{ik} = \sigma_1\sigma_2\sigma_3\,.
\end{aligned}
\qquad (1.30)
$$

We also may define so-called natural invariants S_2 and S_3

$$
\begin{aligned}
S_2 &= \sigma_{ik}\sigma_{ki} = \sigma_1^2 + \sigma_2^2 + \sigma_3^2 \\
S_3 &= \sigma_{ik}\sigma_{kj}\sigma_{ji} = \sigma_1^3 + \sigma_2^3 + \sigma_3^3\,.
\end{aligned}
\qquad (1.31)
$$

Both definitions are linearly related through

$$
\begin{aligned}
S_2^* &= \tfrac{1}{2}(S_2 - S_1^2) \\
S_3^* &= \tfrac{1}{3}[S_3 + \tfrac{1}{2}S_1(S_1^2 - 3S_2)]\,.
\end{aligned}
\qquad (1.32)
$$

The maximum shear stress is

$$|\tau|_{\max} = \frac{1}{2}(\sigma_1 - \sigma_3) \qquad (1.33)$$

when we define $\sigma_1 \geqslant \sigma_2 \geqslant \sigma_3$.

For a plane stress state, we find $S_3^* = 0$ (since $\sigma_3 = 0$, say) and thus the cubic equation (1.29) turns over to a quadratic one

$$\boxed{\sigma^2 - \bar{S}_1\sigma - \tfrac{1}{2}(\bar{S}_2 - \bar{S}_1^2) = 0} \qquad (1.34)$$

with the solutions (1.17).

By defining

$$\boldsymbol{\tau} = \boldsymbol{\sigma} - \sigma_m\mathbf{1}\,, \quad \tau_{ik} = \sigma_{ik} - \sigma_m\delta_{ik}\,, \qquad (1.35)$$

we can introduce a stress deviator $\boldsymbol{\tau}$, such that with the mean stress or hydrostatic stress

$$\sigma_m = \frac{1}{3}S_1\,, \qquad (1.36)$$

the stress tensor can be split into two parts. It is obvious that

$$
\begin{aligned}
T_1 &= \tau_{ii} = 0 \\
T_2 &= \tau_{ik}\tau_{ki} = S_2 - \tfrac{1}{3}S_1^2 \\
T_3 &= \tau_{ik}\tau_{kj}\tau_{ji} = S_3 - S_1 S_2 + \tfrac{2}{9}S_1^3
\end{aligned}
\qquad (1.37)
$$

are the invariants of the stress deviator.

Example 1.1:

An element in plane stress is subjected to stresses $\sigma_{xx} = 160$ MPa, $\sigma_{yy} = 60$ MPa, $\sigma_{xy} = 40$ MPa. Determine: (a) the principal stresses and planes, (b) the stresses on an element rotated through an angle of $45°$, and (c) the maximum shear stresses.

Solution:

(a) To locate the principal planes, we make use of Eq. (1.16), which gives

$$\tan 2\varphi_0 = 0,8 \quad \rightarrow \quad \varphi_0 = 19,33°,$$

From Eq. (1.17), we find

$$\sigma_{1,2} = 110 \pm 64,03; \quad \sigma_1 = 174,03; \quad \sigma_2 = 45,97. \quad \text{[Mpa]}$$

(b) The stresses on an element rotated through an angle of $45°$ can be found from Eq. (1.14). Substituting $2\varphi = 90°$ into these equations gives

$$\sigma_{\bar{x}\bar{x}} = 150, \quad \sigma_{\bar{y}\bar{y}} = 70, \quad \sigma_{\bar{x}\bar{y}} = -50. \quad \text{[Mpa]}$$

(c) The angle of the plane of maximum shear stress is found to be

$$\varphi_1 = 19,33° + 45° = 64,33°, \quad \tau_{\max} = 64,03. \quad \text{[Mpa]}$$

Further, we find

$$\sigma_{\bar{x}\bar{x}} = \sigma_{\bar{y}\bar{y}}|_{\varphi=\varphi_1} = 110. \quad \text{[Mpa]}$$

Example 1.2:

In a problem of plane stress, the normal stresses σ_I, σ_{II} and σ_{III} have been measured in three different directions. Determine:

(a) the principal stresses and planes, and
(b) the maximum shear stresses, with

$$\sigma_I = 100 \text{ MPa}, \quad \sigma_{II} = 50 \text{ MPa},$$
$$\sigma_{III} = -100 \text{ MPa}.$$

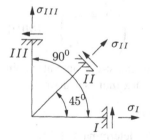

Solution:

Since σ_{xy} has not been measured, we first determine it from Eq. (1.12)$_1$ through

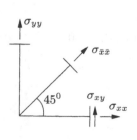

$$\sigma_{\bar{x}\bar{x}} = \frac{1}{2}(\sigma_{xx} + \sigma_{yy}) + \sigma_{xy} \quad \rightarrow$$

$$\sigma_{xy} = 50 \text{ Mpa}$$

With this information, we find from (1.15), and (1.17)

$$\sigma_{1,2} = \pm 111,80 \text{ MPa}, \quad \varphi = 13,15°,$$

$$|\sigma_{\bar{x}\bar{y}}|_{\max} = 111,80 \text{ MPa}, \quad \varphi = 58,15°.$$

Example 1.3:

A plane sheet (plane stress state) is subject
to loads as shown in the figure. Boundary
AC is stress free.

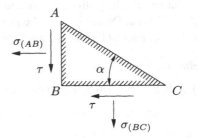

Determine:
(a) the angle α,
(b) the principal stresses and planes,

with: $\sigma_{(AB)} = 40$ MPa, $\sigma_{(BC)} = 20$ MPa

Solution:

From the equilibrium equations, we find

$$\left.\begin{array}{l} \sigma_{(AB)} \sin \alpha + \tau \cos \alpha = 0 \\ \sigma_{(BC)} \cos \alpha + \tau \sin \alpha = 0 \end{array}\right\} \quad \rightarrow \quad \begin{array}{l} \tau^2 = \sigma_{(AB)}\sigma_{(BC)} \\ \tan \alpha = -\sigma_{(BC)}/\tau \end{array}$$

$$\tau = 20\sqrt{2} \text{ MPa}, \quad \alpha = -35, 26°.$$

Then from Eq. (1.15)

$$\sigma_{1,2} = 30 \pm 30, \quad \sigma_1 = 60, \quad \sigma_2 = 0. \quad \text{[MPa]}$$

Example 1.4:

The following matrix of stress components

$$\sigma_{ij} = \begin{pmatrix} 6 & 2 & 2 \\ 2 & 0 & 4 \\ 2 & 4 & 0 \end{pmatrix} \quad \text{[MPa]}$$

describes the stress state at a point of a structure. Determine the principal stresses
and the maximum shear stress.

Solution:

From the determinant (1.27)

$$0 = \begin{vmatrix} 6 - \sigma & 2 & 2 \\ 2 & -\sigma & 4 \\ 2 & 4 & -\sigma \end{vmatrix},$$

we find the cubic equation

$$\sigma^3 - 6\sigma^2 - 24\sigma + 64 = 0 \quad \rightarrow \quad (\sigma + 4)(\sigma - 8)(\sigma - 2) = 0$$

with solutions:

$$\sigma_1 = 8, \quad \sigma_2 = 2, \quad \sigma_3 = -4, \quad \text{[MPa]}$$

$$|\tau|_{max} = \tfrac{1}{2}(\sigma_1 - \sigma_3) = 6. \quad \text{[MPa]}$$

1.3 Strains

During a process of deformation, a specific particle of the body under consideration at point **r** with coordinates (x, y, z) undergoes a displacement **u** with components (u_x, u_y, u_z).

The deformation of a volume element $dx\,dy\,dz$ may be described in various ways. For each axis (and for any other line element), we define here the quotient of its increase in length divided by its original length as its extensional or tensile strain ε. If the volume element is a small cube, its deformation can be described by the tensile strains ε of its three sides and by the changes γ of its originally right angles, the shear strains.

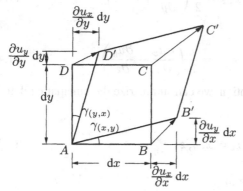

Fig. 1.3
Deformation of a rectangle

Figure 1.3 shows a rectangle $ABCD$ and the quadrilateral $AB'C'D'$ into which it has been deformed during this process. In this figure, the deformed element has been moved back so that its lower left corner is in the position which it had before deformation.

Consequently, the vector BB' has the components

$$\frac{\partial u_x}{\partial x}\,dx, \quad \text{and} \quad \frac{\partial u_y}{\partial x}\,dx,$$

when the partial derivatives of the displacements are small compared to 1. The elongation of the line element AB is

$$AB' - AB = \frac{\partial u_x}{\partial x}\,dx, \tag{1.38}$$

and the strain ε_{xx} of this line element is this quantity divided by the original length $AB = dx$

$$\varepsilon_{xx} = \frac{\partial u_x}{\partial x}. \tag{1.39}$$

Similarly, the strains ε_{yy} and ε_{zz} of the line elements dy and dz are obtained

$$\varepsilon_{yy} = \frac{\partial u_y}{\partial y}, \quad \varepsilon_{zz} = \frac{\partial u_z}{\partial z}. \tag{1.40}$$

In two dimensions, there is only one further strain component, the decrease of the right angle between the sides AB and AD of the element. We define

$$\varepsilon_{xy} = \frac{1}{2}\left(\gamma_{(x,y)} + \gamma_{(y,x)}\right) = \frac{1}{2}\left(\frac{\partial u_y}{\partial x} + \frac{\partial u_x}{\partial y}\right) \tag{1.41}$$

The difference

$$\alpha_{xy} = \frac{1}{2}\left(\frac{\partial u_y}{\partial x} - \frac{\partial u_x}{\partial y}\right) \tag{1.42}$$

defines the average rotation of the element.

In three dimensions, there are two additional shear strains

$$\varepsilon_{xz} = \frac{1}{2}\left(\frac{\partial u_z}{\partial x} + \frac{\partial u_x}{\partial z}\right), \qquad \varepsilon_{yz} = \frac{1}{2}\left(\frac{\partial u_z}{\partial y} + \frac{\partial u_y}{\partial z}\right) \tag{1.43}$$

and rotations

$$\alpha_{xz} = \frac{1}{2}\left(\frac{\partial u_z}{\partial x} - \frac{\partial u_x}{\partial z}\right), \qquad \alpha_{yz} = \frac{1}{2}\left(\frac{\partial u_z}{\partial y} - \frac{\partial u_y}{\partial z}\right). \tag{1.44}$$

Using Einstein's summation convention, we can summarize the kinematic relations (1.39)-(1.41), and (1.43)

$$\boxed{\varepsilon_{ik} = \frac{1}{2}\left(\frac{\partial u_k}{\partial x_i} + \frac{\partial u_i}{\partial x_k}\right)}, \qquad x_i = x, y, z \tag{1.45}$$

From (1.42), (1.44), we find the rotations

$$\boxed{\alpha_{ik} = \frac{1}{2}\left(\frac{\partial u_k}{\partial x_i} - \frac{\partial u_i}{\partial x_k}\right)}, \qquad x_i = x, y, z. \tag{1.46}$$

ε_{ik} are the components of a second-rank tensor, the (infinitesimal) strain tensor, which due to

$$\varepsilon = \varepsilon^T, \qquad \varepsilon_{ik} = \varepsilon_{ki} \tag{1.47}$$

is symmetric. Since

$$\alpha = -\alpha^T, \qquad \alpha_{ik} = -\alpha_{ki}, \qquad \alpha_{ii} = 0, \tag{1.48}$$

α is an antisymmetric deviator.

Sometimes it is more convenient to describe the strains in alternative coordinate systems, such as cylindrical and spherical coordinates:

1. Cylindrical coordinates (r, φ, z)

$$\varepsilon_{rr} = \frac{\partial u_r}{\partial r}, \qquad \varepsilon_{\varphi\varphi} = \frac{u_r}{r} + \frac{1}{r}\frac{\partial u_\varphi}{\partial \varphi}, \qquad \varepsilon_{zz} = \frac{\partial u_z}{\partial z}. \tag{1.49}$$

$$\varepsilon_{r\varphi} = \varepsilon_{\varphi r} = \frac{1}{2} \left\{ \frac{\partial u_\varphi}{\partial r} + \frac{1}{r} \frac{\partial u_r}{\partial \varphi} - \frac{u_\varphi}{r} \right\},$$

$$\varepsilon_{\varphi z} = \varepsilon_{z\varphi} = \frac{1}{2} \left\{ \frac{1}{r} \frac{\partial u_z}{\partial \varphi} + \frac{\partial u_\varphi}{\partial z} \right\}, \tag{1.50}$$

$$\varepsilon_{zr} = \varepsilon_{rz} = \frac{1}{2} \left\{ \frac{\partial u_r}{\partial z} + \frac{\partial u_z}{\partial r} \right\}.$$

2. Spherical coordinates (r, φ, ϑ)

$$\varepsilon_{rr} = \frac{\partial u_r}{\partial r}, \quad \varepsilon_{\varphi\varphi} = \frac{u_r}{r} + \frac{1}{r} \frac{\partial u_\varphi}{\partial \varphi},$$

$$\varepsilon_{\vartheta\vartheta} = \frac{u_r}{r} + \frac{\cot\varphi}{r} u_\varphi + \frac{1}{r\sin\varphi} \frac{\partial u_\vartheta}{\partial \vartheta}. \tag{1.51}$$

$$\varepsilon_{r\varphi} = \varepsilon_{\varphi r} = \frac{1}{2} \left\{ \frac{\partial u_\varphi}{\partial r} + \frac{1}{r} \frac{\partial u_r}{\partial \varphi} - \frac{u_\varphi}{r} \right\},$$

$$\varepsilon_{r\vartheta} = \varepsilon_{\vartheta r} = \frac{1}{2} \left\{ \frac{\partial u_\vartheta}{\partial r} + \frac{1}{r\sin\varphi} \frac{\partial u_r}{\partial \vartheta} - \frac{u_\vartheta}{r} \right\}, \tag{1.52}$$

$$\varepsilon_{\varphi\vartheta} = \varepsilon_{\vartheta\varphi} = \frac{1}{2} \left\{ \frac{1}{r} \frac{\partial u_\vartheta}{\partial \varphi} + \frac{1}{r\sin\varphi} \frac{\partial u_\varphi}{\partial \vartheta} - \frac{\cot\varphi}{r} u_\vartheta \right\}.$$

ε is a symmetric second-rank tensor, and thus follows the same rules, and has the same properties as the stress tensor. This means that we simply take the relations and formulas which have been derived for the stresses and replace herein the stresses by the strains (or any other symmetric second-rank tensor).

Defining the strain deviator

$$\gamma = \varepsilon - \frac{1}{3} e \mathbf{1}, \quad \gamma_{ik} = \varepsilon_{ik} - \frac{1}{3} e \delta_{ik}, \tag{1.53}$$

we can also split up the strain tensor into a first part $e = \varepsilon_{ii}$ describing the volume dilatation, and a second deviator γ_{ik} describing the volume preserving distortions.

1.4 Compatibility Conditions

The kinematic relations (1.45) permit the six strains to be derived from only three displacements u_x, u_y, u_z. Mathematically, this means that the strains cannot all be prescribed arbitrarily as functions of the coordinates; they are connected by a corresponding number of equations. Physically, it means that the strains must be compatible; the deformed elements must fit together.

In three dimensions, there exist 6 compatibility equations

$$\frac{\partial^2 \varepsilon_{ii}}{\partial x_k^2} + \frac{\partial^2 \varepsilon_{kk}}{\partial x_i^2} - 2 \frac{\partial^2 \varepsilon_{ik}}{\partial x_i \partial x_k} = 0, \quad \sum i, k, \tag{1.54}$$

$$\frac{\partial^2 \varepsilon_{ik}}{\partial x_k \partial x_j} + \frac{\partial^2 \varepsilon_{kj}}{\partial x_i \partial x_k} - \frac{\partial^2 \varepsilon_{ij}}{\partial x_k^2} - \frac{\partial^2 \varepsilon_{kk}}{\partial x_i \partial x_j} = 0, \quad \sum k, \tag{1.55}$$

where here the summation convention is excluded for the given indices i and k, and k, respectively.

The six equations cannot all be independent since, if they were, they would be sufficient to determine the six strain components, which, of course, is impossible. It is easily verified that among them three identities hold (Bianchi's identities).

1.5 Equations of Motion

The body under consideration is subject to a sum of forces $\mathbf{F}$ of two kinds; long range and short range. If $\mathcal{B}$ is modeled as a continuum occupying a region V, then the effect of the long-range forces is felt throughout V, while the short-range forces act as contact forces on the boundary surface A.

We start from Newton's second law of motion

$$d\mathbf{F} = d\mathbf{F}_V + d\mathbf{F}_A = \frac{D}{dt}(dm\mathbf{v}) \tag{1.56}$$

which here is given for a volume element $dx\,dy\,dz$ of the body, and where

$$dm = \rho\,dV = \rho\,dx\,dy\,dz \tag{1.57}$$

is the mass of the element, and ρ is the density (mass per unit volume). Thus, any volume element dV experiences a long-range force

$$d\mathbf{F}_V = dm\mathbf{b}, \quad \text{sum of body forces.} \tag{1.58}$$

Any oriented surface element $\mathbf{n}\,dA$ experiences a contact force

$$d\mathbf{F}_A = \mathbf{s}(\mathbf{n})\,dA. \tag{1.59}$$

If $\mathbf{a} = \dot{\mathbf{v}} = \ddot{\mathbf{u}}$ denotes the acceleration field, where $(\bullet) = \dfrac{D(\bullet)}{dt}$ denotes the material time derivative, then the global force equilibrium of motion (balance of linear momentum) is

$$\int_V \rho\,\mathbf{b}\,dV + \int_A \mathbf{s}(\mathbf{n})\,dA = \int_V \rho\,\mathbf{a}\,dV, \tag{1.60}$$

wherein further conservation of mass $\dfrac{D}{dt}(dm) = 0$ is taken into consideration. Equation (1.60) is known as the first of Euler's equations of motion. The second one (balance of angular momentum) has already been used (implicitly) to establish the symmetry of the stress tensor.

With Eq. (1.8) and using Gauss's divergence theorem

$$\int_A \sigma_{ij} n_i \,dA = \int_V \frac{\partial \sigma_{ij}}{\partial x_i}\,dV, \tag{1.61}$$

Eq. (1.60) becomes

$$\int_V \left\{ \frac{\partial \sigma_{ij}}{\partial x_i} + \rho b_j - \rho a_j \right\} dV = 0. \qquad (1.62)$$

This equation embodies a fundamental physical law and thus must be independent of how we define a given body and therefore must be valid for any region V, including very small regions. Consequently, the integrand must be zero, and thus we obtain the local force equation of motion

$$\boxed{\frac{\partial \sigma_{ij}}{\partial x_i} + \rho b_j = \rho a_j} \qquad (1.63)$$

In usual notation, we write

$$\frac{\partial \sigma_{xx}}{\partial x} + \frac{\partial \sigma_{yx}}{\partial y} + \frac{\partial \sigma_{zx}}{\partial z} + \rho b_x = \rho a_x \qquad (1.64)$$

describing the local balance of forces in x direction, and

$$\frac{\partial \sigma_{xy}}{\partial x} + \frac{\partial \sigma_{yy}}{\partial y} + \frac{\partial \sigma_{zy}}{\partial z} + \rho b_y = \rho a_y$$

$$\frac{\partial \sigma_{xz}}{\partial x} + \frac{\partial \sigma_{yz}}{\partial y} + \frac{\partial \sigma_{zz}}{\partial z} + \rho b_z = \rho a_z \qquad (1.65)$$

in y and z directions. For vanishing acceleration terms a_i, Eqs. (1.62) and (1.63) are known as the equilibrium equations of continuum mechanics.

For cylindrical coordinates, we find

$$\frac{\partial \sigma_{rr}}{\partial r} + \frac{1}{r}\frac{\partial \sigma_{\varphi r}}{\partial \varphi} + \frac{\partial \sigma_{zr}}{\partial z} + \frac{1}{r}\left(\sigma_{rr} - \sigma_{\varphi\varphi}\right) + \rho b_r = 0$$

$$\frac{\partial \sigma_{r\varphi}}{\partial r} + \frac{1}{r}\frac{\partial \sigma_{\varphi\varphi}}{\partial \varphi} + \frac{\partial \sigma_{z\varphi}}{\partial z} + \frac{2}{r}\sigma_{r\varphi} + \rho b_\varphi = 0 \qquad (1.66)$$

$$\frac{\partial \sigma_{rz}}{\partial r} + \frac{1}{r}\frac{\partial \sigma_{\varphi z}}{\partial \varphi} + \frac{\partial \sigma_{zz}}{\partial z} + \frac{1}{r}\sigma_{rz} + \rho b_z = 0.$$

The corresponding equations in spherical coordinates are

$$\frac{\partial \sigma_{rr}}{\partial r} + \frac{1}{r}\frac{\partial \sigma_{\varphi r}}{\partial \varphi} + \frac{1}{r\sin\varphi}\frac{\partial \sigma_{\vartheta r}}{\partial \vartheta} +$$

$$+ \frac{1}{r}\left(2\sigma_{rr} - \sigma_{\varphi\varphi} - \sigma_{\vartheta\vartheta} + \sigma_{\vartheta r}\right) + \rho b_r = 0$$

$$\frac{\partial \sigma_{\varphi\varphi}}{\partial \varphi} + \frac{1}{r\sin\varphi}\frac{\partial \sigma_{\vartheta\varphi}}{\partial \vartheta} + \frac{1}{r}\left(\sigma_{\varphi\varphi}\cot\varphi - \sigma_{\vartheta\vartheta}\cot\varphi + 3\sigma_{r\varphi}\right) + \rho b_\varphi = 0$$

$$\frac{\partial \sigma_{r\vartheta}}{\partial r} + \frac{1}{r}\frac{\partial \sigma_{\varphi\vartheta}}{\partial \varphi} + \frac{1}{r\sin\varphi}\frac{\partial \sigma_{\vartheta\vartheta}}{\partial \vartheta} + \frac{1}{r}\left(3\sigma_{r\vartheta} + 2\sigma_{\varphi\vartheta}\cot\varphi\right) + \rho b_\vartheta = 0.$$

$$(1.67)$$

1.6 Energy

We now consider our body in a state of motion. A volume element dV of this body in a time interval dt experiences an infinitesimal additional displacement

$$\mathbf{du} = \mathbf{v}\, dt. \tag{1.68}$$

To determine the energy balance, we again start with Newton's second law of motion (Eq. 1.56) and multiply this equation by $\mathbf{du}$. The global balance thus becomes

$$\int_V d\mathbf{F} \cdot \mathbf{du} = \int_V dm\, \dot{\mathbf{v}} \cdot \mathbf{du} = \int_V \rho\, \dot{\mathbf{v}} \cdot \mathbf{du}\, dV. \tag{1.69}$$

Using the divergence theorem, we find – in index notation –

$$\int_V \left\{ \frac{\partial \sigma_{ik}}{\partial x_i} + \rho\, b_k \right\} du_k\, dV = \int_V \rho\, \dot{v}_k\, du_k\, dV \tag{1.70}$$

and further

$$\int_V \left\{ \frac{\partial}{\partial x_i}(\sigma_{ik}\, du_k) - \sigma_{ik} \frac{\partial}{\partial x_i}(du_k) + \rho\, b_k\, du_k \right\} dV = \frac{1}{2} \int_V \rho\, D(v_k^2)\, dV. \tag{1.71}$$

The first term gives (again using the divergence theorem)

$$\int_V \frac{\partial}{\partial x_i}(\sigma_{ik}\, du_k)\, dV = \int_A s_k\, du_k\, dA = dA_A \tag{1.72}$$

the increment of the external work of all short-range forces.

Apparently, the third term describes the incremental work of the long-range forces

$$\int_V \rho\, b_k\, du_k\, dV = dA_V. \tag{1.73}$$

Due to the symmetry of the stresses, the second term is

$$\int_V \sigma_{ik} \frac{\partial}{\partial x_i}(du_k)\, dV = \int_V \sigma_{ik}\, d\varepsilon_{ik}\, dV = dW, \tag{1.74}$$

the increment of the internal work or energy related with the deformation of the body. Thus for rigid bodies this term must vanish.

Finally, the right-hand side of Eq. (1.69) gives the increment of the kinetic energy

$$\frac{1}{2} \int_V \rho\, D(v_k^2)\, dV = dE. \tag{1.75}$$

Thus, we find

$$\boxed{dA = dA_A + dA_V = dW + dE,} \tag{1.76}$$

the energy balance of mechanics. Any change of the work dA done by the short-range and long-range forces causes a change of either the internal work dW or the kinetic energy dE.

For static problems, with $dE = 0$, this means that any change of dA causes a change of dW, or simply

$$dW - dA = d(W - A) = 0, \qquad (1.77)$$

expressing the stationarity of the function $W - A$. This means that for a system in equilibrium the function $W - A$ attains a stationary value.

1.7 Principle of Virtual Work

The concepts of virtual displacements and virtual work are usually introduced during the study of statics, where they are used to solve problems of static equilibrium. The word "virtual" implies that the quantities are purely hypothetical and that they do not exist in a real physical sense. Thus a virtual displacement is an imaginary (infinitesimal) continuous displacement satisfying the prescribed geometrical boundary conditions, arbitrarily imposed upon the system. The work done by the real external and internal forces during a virtual displacement is called virtual work.

Similar as before, we introduce virtual displacements δu_i - in index notation - which are related to virtual strains $\delta \varepsilon_{ik}$ by

$$\delta \varepsilon_{ik} = \frac{1}{2} \left\{ \frac{\partial \delta u_k}{\partial x_i} + \frac{\partial \delta u_i}{\partial x_k} \right\}. \qquad (1.78)$$

As in the preceding Section, we start by multiplying the equilibrium equations by δu_k and thus arrive at

$$\int_V \left\{ \frac{\partial \sigma_{ik}}{\partial x_i} + \rho b_k \right\} \delta u_k \, dV = 0 \qquad (1.79)$$

and further

$$\int_V \left\{ \frac{\partial}{\partial x_i} (\sigma_{ik} \delta u_k) - \sigma_{ik} \frac{\partial}{\partial x_i} (\delta u_k) + \rho b_k \delta u_k \right\} dV = 0. \qquad (1.80)$$

The first term gives (again using the divergence theorem)

$$\int_V \frac{\partial}{\partial x_i} (\sigma_{ik} \delta u_k) \, dV = \int_A s_k \, \delta u_k \, dA = \delta A_A, \qquad (1.81)$$

the virtual work of the external forces, prescribed on the surface of the body.

It is easy to see that the third term describes the virtual work of the long-range forces

$$\int_V \rho b_k \delta u_k \, dV = \delta A_V. \qquad (1.82)$$

Due to the symmetry of the stresses, the second term is

$$\int_V \sigma_{ik} \frac{\partial}{\partial x_i} (\delta u_k) \, dV = \int_V \sigma_{ik} \delta \varepsilon_{ik} \, dV = \delta W, \tag{1.83}$$

the virtual work of the internal forces. The principle of virtual work may thus be stated

$$0 = -\int_V \sigma_{ik} \, \delta \varepsilon_{ik} \, dV + \int_V b_k \, \delta u_k \rho \, dV + \int_A s_k \, \delta u_k \, dA$$

$$= -\delta W + \delta A_V + \delta A_A \,, \, \cdot \tag{1.84}$$

similar to Eq. (1.75), and finally

$$\delta W - \delta A = \delta (W - A) = 0. \tag{1.85}$$

We note that this result is independent of any constitutive relation.

The principle, however, may also be stated alternatively in the following manner: If $\delta(W - A)$ vanishes for any arbitrary infinitesimal virtual displacements satisfying the prescribed geometrical constraints, the mechanical system is in equilibrium. Thus, the principle of virtual work is equivalent to the equations of equilibrium of the system. However, the former has a much wider field of application to the formulation of mechanics problems than the latter.

1.8 Exercises to Chapter 1

Problem 1.1:

For the stress tensor

$$\boldsymbol{\sigma} = \begin{pmatrix} \sigma_{xx} & \sigma_{xy} & \sigma_{xz} \\ \sigma_{yx} & \sigma_{yy} & \sigma_{yz} \\ \sigma_{zx} & \sigma_{zy} & \sigma_{zz} \end{pmatrix},$$

compute the natural (basic) invariants

$$S_1 = \text{tr} \, [\boldsymbol{\sigma}] = \sigma_{ii}, \quad S_2 = \text{tr} \, [\boldsymbol{\sigma}^2], \quad S_3 = \text{tr} \, [\boldsymbol{\sigma}^3]$$

and the principal invariants

$$S_1 = S_1, \quad S_2^* = \frac{1}{2}(S_2 - S_1^2), \quad S_3^* = \det \boldsymbol{\sigma} = \frac{1}{6}(S_1^3 - 3S_1 S_2 + 2S_3).$$

Give the principal invariants T_1, T_2^*, T_3^* of the stress deviator

$$\boldsymbol{\tau} = \boldsymbol{\sigma} - \sigma_m \mathbf{1}, \quad \sigma_m = \frac{1}{3} S_1$$

as functions of the principal invariants of the stress tensor $\boldsymbol{\sigma}$. Show that $\boldsymbol{\tau}$ and $\boldsymbol{\sigma}$ have the same principal directions and that the eigenvalues τ_i of $\boldsymbol{\tau}$ are related to the eigenvalues σ_i of $\boldsymbol{\sigma}$ by the expression

$T_i = \sigma_i - \sigma_m.$

Use the substitution

$$\tau = 2\sqrt{\frac{1}{3}\, T_2^*}\ \cos\varphi$$

to find the roots of the characteristic equation

$$\tau^3 - T_2^*\tau - T_3^* = 0$$

of the stress deviator, and hence show that the eigenvalues of σ are

$$\sigma_i = \frac{1}{3}\left(S_1 + 2\sqrt{S_1^2 + 3S_2^*}\ \cos\varphi_i\right), \qquad (i = 1, 2, 3)$$

with

$$\varphi_1 = \frac{1}{3}\cos^{-1}\left[\frac{27S_3^* + 9S_2^*S_1 + 2S_1^3}{2\left(3S_2^* + S_1^2\right)^{\frac{3}{2}}}\right], \quad \varphi_2 = \varphi_1 + \frac{2\pi}{3}, \quad \varphi_3 = \varphi_1 - \frac{2\pi}{3}.$$

Problem 1.2:

Use the results of Problem 1.1 to compute the eigenvalues of the following (stress and strain) tensors:

$$\sigma_a = \begin{pmatrix} 9 & -12 & -3 \\ -12 & 18 & -12 \\ -3 & -12 & 9 \end{pmatrix}, \quad \sigma_b = \begin{pmatrix} \frac{35}{2} & 5 & -\frac{5}{2} \\ 5 & 20 & 5 \\ -\frac{5}{2} & 5 & \frac{35}{2} \end{pmatrix},$$

$$\sigma_c = \begin{pmatrix} \frac{72}{5} & 0 & \frac{96}{5} \\ 0 & 40 & 0 \\ \frac{96}{5} & 0 & \frac{128}{5} \end{pmatrix}, \quad \varepsilon_a = \begin{pmatrix} 1 & 0 & 0 \\ 0 & 0 & -2 \\ 0 & -2 & 3 \end{pmatrix} \cdot 10^{-4},$$

$$\varepsilon_b = \begin{pmatrix} 2 & -1 & 1 \\ -1 & 0 & 1 \\ 1 & 1 & 2 \end{pmatrix} \cdot 10^{-4}, \quad \varepsilon_c = \begin{pmatrix} 0 & -1 & 1 \\ -1 & 0 & 1 \\ 1 & 1 & 0 \end{pmatrix} \cdot 10^{-4}.$$

Problem 1.3:

For an element in plane stress subjected to the stresses $\sigma_{xx} = 150$, $\sigma_{xy} = 80$, and $\sigma_{yy} = -50$, find using the transformation equations

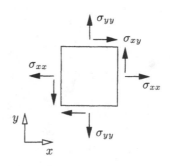

a) the principal stresses and planes,
b) the stresses on an element rotated by an angle of $45°$,
c) the maximum shear stresses.

Determine these results also using Mohr's circle.

Problem 1.4:

Compute for the plane deformation field

$$u_x = \frac{x}{l_0} + \frac{3y}{l_0}, \quad u_y = \frac{3x}{l_0} - \frac{3y}{l_0}$$

1. the strain tensor ε,
2. the rotation tensor α, and
3. the principal values and directions of ε.

Problem 1.5:

For the 60° strain rosette shown in the figure, gage A measures the normal strain ε_a
in the direction of the x axis. Gages B
and C measure the strains ε_b and ε_c, respectively. Compute the strain tensor ε_{ik},
its principal invariants, and the principal
strains ε_1 and ε_2.

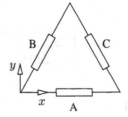

2. Elastic Material

2.1 Hooke's Law

The preceding discussion of stress and strain is concerned with kinematics and statics, and the relations obtained are valid for any deformable medium, solid or fluid. However, the kinematic relations (1.45) and the equilibrium conditions (Eqs. 1.63, with vanishing accelerations) give only 9 equations for 15 unknowns - 3 displacements, 6 strains, and 6 stresses - and, hence, are insufficient to determine these quantities. The third set of equations needed consists of relations between stress and strain, which must be found from experiment, and thus depends highly on the material. According to their stress-strain relations, we distinguish between elastic, plastic, viscoelastic solids, viscous and ideal fluids, etc.

The theory of (isotropic) elasticity is built on Hooke's law, which involves the following assumptions about the properties of the material:

(i) proportionality of stress and strain,
(ii) elasticity, i.e. the capability of storing energy which can be recovered without loss,
(iii) isotropy, i.e. independence of the elastic properties of the direction of the stress,
(iv) homogeneity, i.e. independence of these properties of the coordinates of the point considered.

In its original form, Hooke's law was formulated as an anagram, stating a linear relation between action (stress) and reaction (strain)

$$\sigma = E\varepsilon \tag{2.1}$$

for the description of a one-dimensional problem. It may be easily verified that the most general law which corresponds to (i) and (iii) is

$$
\begin{aligned}
E\varepsilon_1 &= \sigma_1 - \nu(\sigma_2 + \sigma_3), \\
E\varepsilon_2 &= \sigma_2 - \nu(\sigma_3 + \sigma_1), \\
E\varepsilon_3 &= \sigma_3 - \nu(\sigma_1 + \sigma_2).
\end{aligned} \tag{2.2}
$$

The modulus of elasticity E (Young's modulus) and Poisson's ratio ν must be determined from experiment. E has the dimension of a stress, and varies widely between different materials. Poisson's ratio is confined to lie between 0 and 0.5. According to (ii) E and ν are independent of time; according to (iv) they are independent of r.

Hooke's law may be further transformed to an arbitrary orthogonal frame. The most general linear relation is - in index notation -

$$\sigma_{ij} = C_{ijkl}\,\varepsilon_{kl}\,, \tag{2.3}$$

where C_{ijkl} is a fourth-rank linear material tensor with $3^4 = 81$ coordinates. Some symmetry properties reduce this number to 21 independent material parameters for general anisotropic linear elastic materials (6 independent parameters, respectively, for plane stress state).

Equation (2.2) may uniquely be inverted to give

$$\varepsilon_{ij} = S_{ijkl}\,\sigma_{kl}\,, \tag{2.4}$$

where the compliance tensor

$$S_{ijkl} = (C_{ijkl})^{-1} \tag{2.5}$$

is the inverse of the elasticity tensor.

For an isotropic, and linear elastic material, the number of independent material parameters may be further reduced to two. We get

$$\sigma_{ik} = 2G\left(\varepsilon_{ik} + \frac{\nu}{1-2\nu}\,e\,\delta_{ik}\right), \tag{2.6}$$

and the inverse

$$\varepsilon_{ik} = \frac{1}{2G}\left(\sigma_{ik} - \frac{3\nu}{1+\nu}\,\sigma_m\delta_{ik}\right), \tag{2.7}$$

where the shear modulus G is given by

$$G = \frac{E}{2(1+\nu)}\,. \tag{2.8}$$

When stress and strain tensor are split into deviatoric and hydrostatic parts, Hooke's law may be written in the particularly simple form

$$\sigma_m = K\,e\,, \quad \tau_{ik} = 2G\,\gamma_{ik}\,, \tag{2.9}$$

where

$$K = \frac{E}{3(1-2\nu)} \tag{2.10}$$

is the bulk modulus, and e is the volume dilatation. We emphasize that with Eqs. (2.9) the volumetric (dilatational) and deviatoric (distortional) parts of the stress-strain relations are uncoupled.

Although E and ν, and thus the shear modulus G, are measured experimentally more easily than others, these parameters are sometimes replaced by the Lamé coefficients λ and μ, e.g. the elasticity tensor C_{ijkl} (for an isotropic material) takes the following form

$$C_{ijkl} = \mu(\delta_{ik}\delta_{jl} + \delta_{il}\delta_{jk}) + \lambda\,\delta_{ij}\delta_{kl}\,. \tag{2.11}$$

Introducing this expression into Eq. (2.3), constitutive relation (2.6) may be replaced by the somewhat simpler form

$$\sigma_{ik} = 2\mu\varepsilon_{ik} + \lambda e\,\delta_{ik}\,, \quad \boldsymbol{\sigma} = 2\mu\boldsymbol{\varepsilon} + \lambda e\,\mathbf{1}\,, \tag{2.12}$$

where the following identities hold

$$E = \frac{\mu(3\lambda + 2\mu)}{\lambda + \mu}\,, \quad \nu = \frac{\lambda}{2(\lambda + \mu)}\,. \tag{2.13}$$

The Lamé coefficients are easily deduced

$$\lambda = \frac{E\nu}{(1 - 2\nu)(1 + \nu)}\,, \quad \mu = \frac{E}{2(1 + \nu)} = G\,. \tag{2.14}$$

In the sequel, we prefer to use the material parameters E and ν, and G, respectively, rather than the Lamé coefficients.

2.2 Strain Energy, Complementary Energy

In Section 1.6, we introduced the increment of internal energy dW as

$$dW = \int_V dw\,dm = \int_V \frac{1}{\rho}\,\sigma_{ik}\,d\varepsilon_{ik}\,dm\,. \tag{2.15}$$

The local quantity herein

$$dw = \frac{1}{\rho}\,\sigma_{ik}\,d\varepsilon_{ik} \tag{2.16}$$

will be called increment of the specific internal energy, i.e. the increment of the internal energy per unit mass.

We can now introduce into this relation the additive split of both tensors σ_{ik} and $d\varepsilon_{ik}$ into their deviatoric and hydrostatic (volumetric) parts and obtain

$$dw = dw_V + dw_G = \frac{1}{\rho}\,\sigma_m\,de + \frac{1}{\rho}\,\tau_{ik}\,d\gamma_{ik}\,, \tag{2.17}$$

where the first term dw_V describes the energy (work) related with the volumetric (dilatational) changes, whereas dw_G - on the other hand - is related to the shear (distortional) deformations.

If we, further, suppose any initial state as a "natural state", i.e. that at time $t = t_0$: $\sigma_{ik}|_{t_0} = 0$, $\varepsilon_{ik}|_{t_0} = 0$ is assumed, then using the generalized Hooke's law (2.6), Eq. (2.17) may be integrated to give

$$\rho w(\varepsilon_{ik}) = \frac{1}{2}\,Ke^2 + G\gamma_{ik}\gamma_{ik}\,. \tag{2.18}$$

It may easily be shown that this specific strain energy can also be written as

$$\rho w(\varepsilon_{ik}) = G\left(\varepsilon_{ik}\varepsilon_{ik} + \frac{\nu}{1 - 2\nu}\,e^2\right)\,. \tag{2.19}$$

Using Eqs. (2.9) and (2.7), this may also be expressed as a function of the stresses

$$\rho w(\sigma_{ik}) = \rho w^*(\sigma_{ik}) = \frac{1}{2K}\, \sigma_m^2 + \frac{1}{4G}\, \tau_{ik}\tau_{ik} \tag{2.20}$$

and

$$\rho w(\sigma_{ik}) = \rho w^*(\sigma_{ik}) = \frac{1}{4G}\left(\sigma_{ik}\sigma_{ik} - \frac{9\nu}{1+\nu}\, \sigma_m^2\right). \tag{2.21}$$

Since the sum of Eqs. (2.21) and (2.19) (or (2.20) and (2.18)) gives

$$w(\varepsilon_{ik}) + w^*(\sigma_{ik}) = \frac{1}{\rho}\, \sigma_{ik}\varepsilon_{ik}, \tag{2.22}$$

the latter is called specific complementary energy.

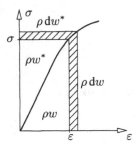

Fig. 2.1
Strain and complementary energies in uniaxial tension

Introducing here in analogy with Eq. (2.16)

$$dw^* = \frac{1}{\rho}\, \varepsilon_{ik}\, d\sigma_{ik}, \tag{2.23}$$

the increment of the specific complementary energy, we see immediately that Eq. (2.22) also holds for a nonlinear elastic material (provided our considerations are restricted to small deformations with $\rho \cong \rho_0$). This is illustrated in Fig. 2.1 in the uniaxial case by the marked areas below and above the stress-strain curve. It is seen that $\rho w(\varepsilon_{ik})$ and $\rho w^*(\sigma_{ik})$ are complementary to each other in representing the total area $\sigma \cdot \varepsilon$.

2.3 Fundamental Equations

In addition to the field equations, a set of boundary conditions is needed: either the displacements u (or derivatives of its components) or the surface tractions s must be prescribed on the surface of the body.

A reduction of the 15 field equations can be accomplished in various ways, two of which will be mentioned here:

(i) The first of the reduced systems is obtained, when we introduce the kinematic relations and the stress-strain relations into the equilibrium equations.

This yields three equations

$$\Delta u_i + \frac{1}{1 - 2\nu} \frac{\partial e}{\partial x_i} + \rho \frac{b_i}{G} = 0, \tag{2.24}$$

where

$$e = \varepsilon_{ii} = \frac{\partial u_i}{\partial x_i} \tag{2.25}$$

is the dilatation, and

$$\Delta = \frac{\partial^2}{\partial x^2} + \frac{\partial^2}{\partial y^2} + \frac{\partial^2}{\partial z^2} \tag{2.26}$$

is the three-dimensional Laplace operator. These are the fundamental equations of elasticity (or the Lamé-Navier's equations).

(ii) The second system of equations are the six Beltrami's equations for the stresses.

$$\Delta \sigma_{ik} + \frac{3}{1 + \nu} \frac{\partial^2 \sigma_m}{\partial x_i \partial x_k} + \rho \left(\frac{\nu}{1 - \nu} \frac{\partial b_l}{\partial x_l} \delta_{ik} + \frac{\partial b_i}{\partial x_k} + \frac{\partial b_k}{\partial x_i} \right) = 0, \tag{2.27}$$

where

$$\sigma_m = \frac{1}{3} \sigma_{ii} \tag{2.28}$$

is the mean stress (hydrostatic stress). Equation (2.27) is obtained by substituting the stress-strain relation into the compatibility conditions.

These are two different sets of mathematically similar equations for solving the field problem of linear isothermal elasticity. Which way of solution we prefer is mainly related with the given boundary conditions. For problems with prescribed displacements the first way would be advantageous, whereas for prescribed tractions the second way could be more favourable.

2.4 Influence of Temperature

A uniform, i.e. homogeneous temperature field causes a volumetric change of the body under consideration. For the different materials, we find

$$\varepsilon_{ik} = \frac{1}{2G} \left(\sigma_{ik} - \frac{3\nu}{1 + \nu} \sigma_m \delta_{ik} \right) + \alpha \Theta \, \delta_{ik}, \tag{2.29}$$

where α is the linear expansion coefficient of the material, and Θ is the increase in temperature.

The inverse of Eq. (2.29) is

$$\sigma_{ik} = 2G \left(\varepsilon_{ik} + \frac{\nu}{1 - 2\nu} e \, \delta_{ik} \right) - \frac{E}{1 - 2\nu} \alpha \Theta \, \delta_{ik}, \tag{2.30}$$

and, consequently, Eq. $(2.9)_1$ changes to

$$\sigma_m = K(e - 3\alpha\Theta), \tag{2.31}$$

whereas the deviatoric part $(2.9)_2$ remains unchanged.

For a rectangular cartesian coordinate system, we have (Eq. 2.29)

$$\varepsilon_{xx} = \frac{1}{E}\left\{\sigma_{xx} - \nu(\sigma_{yy} + \sigma_{zz})\right\} + \alpha\Theta$$

$$\varepsilon_{yy} = \frac{1}{E}\left\{\sigma_{yy} - \nu(\sigma_{zz} + \sigma_{xx})\right\} + \alpha\Theta$$

$$\varepsilon_{zz} = \frac{1}{E}\left\{\sigma_{zz} - \nu(\sigma_{xx} + \sigma_{yy})\right\} + \alpha\Theta$$

$$\varepsilon_{xy} = \frac{1}{2G}\,\sigma_{xy}$$ \hfill (2.32)

$$\varepsilon_{yz} = \frac{1}{2G}\,\sigma_{yz}$$

$$\varepsilon_{zx} = \frac{1}{2G}\,\sigma_{zx}\,,$$

and from the inverse relation (Eq. 2.30)

$$\sigma_{xx} = \frac{2G}{1-2\nu}\left\{(1-\nu)\varepsilon_{xx} + \nu(\varepsilon_{yy} + \varepsilon_{zz}) - (1+\nu)\alpha\Theta\right\}$$

$$\sigma_{yy} = \frac{2G}{1-2\nu}\left\{(1-\nu)\varepsilon_{yy} + \nu(\varepsilon_{zz} + \varepsilon_{xx}) - (1+\nu)\alpha\Theta\right\}$$

$$\sigma_{zz} = \frac{2G}{1-2\nu}\left\{(1-\nu)\varepsilon_{zz} + \nu(\varepsilon_{xx} + \varepsilon_{yy}) - (1+\nu)\alpha\Theta\right\}$$ \hfill (2.33)

$$\sigma_{xy} = 2G\,\varepsilon_{xy}$$

$$\sigma_{yz} = 2G\,\varepsilon_{yz}$$

$$\sigma_{zx} = 2G\,\varepsilon_{zx}\,.$$

2.5 Hooke's Law in Two Dimensions

There are two ways of specializing Eqs. (2.32) and (2.33) to two dimensions, letting either $\sigma_{iz} = 0$ (plane stress) or $\varepsilon_{iz} = 0$ (plane strain). The equations relating stress and strain, in each case, reduce accordingly:

1. Plane stress:

$$\sigma_{zx} = \sigma_{zy} = \sigma_{zz} = 0.$$

From Eq. (2.32), we get

$$\varepsilon_{xx} = \frac{1}{E}(\sigma_{xx} - \nu\sigma_{yy}) + \alpha\Theta$$

$$\varepsilon_{yy} = \frac{1}{E}(\sigma_{yy} - \nu\sigma_{xx}) + \alpha\Theta$$ \hfill (2.34)

$$\varepsilon_{xy} = \frac{1}{2G}\,\sigma_{xy}$$

and, further,

$$\varepsilon_{zz} = \frac{-\nu}{E}\left(\sigma_{xx} + \sigma_{yy}\right) + \alpha\Theta \tag{2.35}$$

in the z direction.

2. Plane strain:

$$\varepsilon_{zx} = \varepsilon_{zy} = \varepsilon_{zz} = 0.$$

From Eq. (2.32), we get

$$\varepsilon_{xx} = \frac{1}{2G}\left\{(1-\nu)\sigma_{xx} - \nu\sigma_{yy}\right\} + (1+\nu)\alpha\Theta$$

$$\varepsilon_{yy} = \frac{1}{2G}\left\{(1-\nu)\sigma_{yy} - \nu\sigma_{xx}\right\} + (1+\nu)\alpha\Theta \tag{2.36}$$

$$\varepsilon_{xy} = \frac{1}{2G}\sigma_{xy}$$

and in the z direction

$$\sigma_{zz} = \nu(\sigma_{xx} + \sigma_{yy}) - E\alpha\Theta. \tag{2.37}$$

The problems of plane stress and plane strain, although different in their basic equations, are not independent of each other. When a plane stress problem has been solved, the solution for the corresponding plane strain problem may be derived from it by the following procedure:

Replace

$$E \Rightarrow \frac{E'}{1-\nu'^2}, \quad \nu \Rightarrow \frac{\nu'}{1-\nu'}, \quad \alpha \Rightarrow \alpha'(1+\nu') \tag{2.38}$$

and then drop the primes.

Conversely, if the plane strain problem has been solved the solution for plane stress is obtained by the procedure:

Replace

$$E \Rightarrow \frac{E'(1+2\nu')}{(1+\nu')^2}, \quad \nu \Rightarrow \frac{\nu'}{1+\nu'}, \quad \alpha \Rightarrow \alpha'\frac{1+\nu'}{1+2\nu'} \tag{2.39}$$

and then drop the primes.

It is easily seen that neither substitution changes the shear modulus G.

Example 2.1:

A reinforced concrete plate has been built at a temperature of $+5°C$. What happens if the plate is heated to a constant temperature of $55°C$?

Determine the stresses in the plate, and the horizontal forces in the supports if thermal expansion is restrained:

$$E = 2.1 \cdot 10^4 \text{ MPa}, \quad \alpha = 10^{-5} \text{ K}^{-1}$$

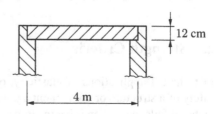

Solution:

We introduce a cartesian coordinate system with the x-axis in horizontal direction, and the y-axis in thickness direction. From Hooke's law, we find (Eq. 2.34)

$$\varepsilon_{xx} = \frac{1}{E}\sigma_{xx} + \alpha\Theta$$

for a plane stress state, and a vanishing stress σ_{yy}.

For a restrained expansion, i.e. $\varepsilon_{xx} = 0$, we determine

$$\sigma_{xx} = -E\alpha\Theta = -10.5\,\text{MPa} \quad \rightarrow \quad f = -1260\,\text{kN/m}.$$

Analogously, for a plane strain situation (Eq. 2.36), with $\nu = 0.3$

$$\sigma_{xx} = \sigma_{zz} = -\frac{E}{1-\nu}\alpha\Theta = -15\,\text{MPa} \quad \rightarrow \quad f = -1800\,\text{kN/m}.$$

Example 2.2:

A quadratic piece of rubber with dimension a is pressed into a quadratic hole. The stamp and the hole may be assumed rigid.

Determine the stresses in the rubber. What happens if Poisson's ratio ν tends to 0.5?

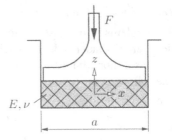

Solution:

Due to the assumed rigidity, we find

$$\varepsilon_{xx} = \varepsilon_{yy} = 0, \quad \rightarrow \quad \sigma_{xx} = \sigma_{yy}.$$

Introducing $\sigma_{zz} = -F/a^2$, we determine from Eqs. (2.32)

$$\sigma_{xx} = \sigma_{yy} = -\frac{\nu}{1-\nu}\frac{F}{a^2},$$

and furthermore

$$\varepsilon_{zz} = -\frac{(1+\nu)(1-2\nu)}{1-\nu}\frac{F}{Ea^2}.$$

Thus, for $\nu = 0.5$ the material behaves like a rigid body.

2.6 Strength Criteria

In technical applications of elasticity, one important objective is to ascertain the safety of a structure or component against failure. Under uniaxial stress, a metallic material fails, e.g. when it begins to deform plastically, i.e. when the yield stress is reached. This situation can be avoided with a judicious choice of a safety factor in deciding on a safe working elastic stress for the material. However, in practice, the stress in loaded structures is often multiaxial. The question then arises, what magnitudes of these combined stresses will cause the onset of yielding. It is necessary to find a suitable criterion based upon stress, strain or energy for the complex system that can be related to the corresponding quantity at the uniaxial yield stress σ_0, which is most conveniently measured from a tension test.

For an isotropic elastic material this criterion must be independent from any coordinate system in which stress or strain components have been determined. Thus, a reasonable criterion must be based on the invariants of stresses and strains.

In what follows, only a few of them will be discussed, since now it is recognized that those attributed to v. Mises and Tresca are the most representative to describe the behaviour of (ductile) metallic materials.

(i) Maximum Principal Stress Theory (Rankine)

The simplest criterion states that yielding commences either when the major principal stress σ_1 attains the value of the tensile yield stress σ_0, or when the minor principal stress σ_3, if $|\sigma_3| > |\sigma_1|$, reaches $-\sigma_0$. That is,

$$\sigma_{eq} = \sigma_1, \quad \text{or} \quad \sigma_{eq} = -\sigma_3. \tag{2.40}$$

These equations ignore the intermediate principal stress (σ_2) and assume, along with all other criteria, that the tensile and compressive yield stress are equal.

(ii) Maximum Shear Stress Theory (Coulomb, Tresca, Guest)

This often used theory assumes that yielding begins when the maximum shear stress reaches a critical value. In the simple tension test with $\sigma_1 = \sigma$, $\sigma_2 = \sigma_3 = 0$, we have

$$|\tau|_{max} = \frac{1}{2}\sigma. \tag{2.41}$$

Equating the shear stresses for the uniaxial and multiaxial cases, leads to the Tresca criterion

$$\sigma_{eq} = 2|\tau|_{max} = \sigma_1 - \sigma_3. \tag{2.42}$$

(iii) Total Strain Energy Theory (Beltrami, Haigh)

The theory assumes that yielding commences when the total internal energy stored attains the value of the internal energy for uniaxial yielding. Thus from (2.21)

$$\rho w^*(\sigma_{ik}) = \frac{1}{4G}\left(S_2 - \frac{\nu}{1+\nu}S_1^2\right), \tag{2.43}$$

expressed in terms of invariants of the stresses. In the uniaxial case ($\sigma_1 = \sigma$, $\sigma_2 = \sigma_3 = 0$), we find

$$\rho w^*(\sigma) = \frac{1}{2E}\sigma^2, \tag{2.44}$$

and thus, equating this expression with the energy for the multiaxial case, we find

$$\sigma_{eq} = \sqrt{(1+\nu)S_2 - \nu S_1^2}$$

$$= \sqrt{\sigma_1^2 + \sigma_2^2 + \sigma_3^2 - 2\nu(\sigma_1\sigma_2 + \sigma_2\sigma_3 + \sigma_3\sigma_1)},$$

(2.45)

expressed in terms of principal stresses.

(iv) Shear Strain Energy Theory (Maxwell, Huber, v. Mises, Hencky)

This theory assumes that only the distortional part of the internal energy will contribute to yielding, and thus from the second term of (2.20), we find

$$\rho w^*(\sigma_{ik}) = \frac{1}{4G} T_2,$$

(2.46)

expressed in terms of the second invariant of the stress deviator. From the uniaxial case follows $(1.34)_2$

$$\rho w^*(\sigma) = \frac{1}{4G}\frac{2}{3}\sigma^2.$$

(2.47)

Again, equating this expression with the distortional part of the internal energy for multiaxial problems yields

$$\sigma_{eq} = \sqrt{\frac{3}{2}T_2} = \sqrt{\sigma_1^2 + \sigma_2^2 + \sigma_3^2 - \sigma_1\sigma_2 - \sigma_2\sigma_3 - \sigma_3\sigma_1},$$

(2.48)

the well-known v. Mises criterion. The four criteria may be compared graphically

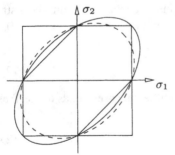

Fig. 2.2
Different strength criteria in plane stress

in principal axes σ_1 and σ_2, when the third principal stress $\sigma_3 = 0$ (see Fig. 2.2). Clearly, the energy criteria both describe ellipses with a $45°$ inclination in their major axes (ν in criterion (iii) - dashed line - is here taken to be 0.3). Putting $\sigma_3 = 0$ in Eqs. (2.40) and (2.42) results in a square and a hexagon shown when each respective criterion is applied separately to the stress state existing within each quadrant.

Example 2.3:
A closed thin-walled tube (container) with: radius R, thickness H, is subjected to internal pressure p, and additional torsion with shear stress τ (a plane stress state is supposed). The stresses in a cylindrical coordinate system are then

$$\sigma_{\varphi\varphi} = \sigma_\varphi = p\frac{R}{H}, \quad \sigma_{zz} = \sigma_z = \frac{1}{2}\sigma_\varphi, \quad \sigma_{\varphi z} = \tau.$$

Determine the equivalent stresses according to the different criteria:

(a) Maximum principal stress, (b) maximum shear stress, (c) total strain energy, and (d) shear strain energy.

Solution:

The principal stresses are (Eq. 1.17)

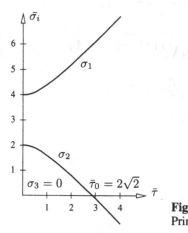

Fig. 2.3
Principal stresses vs. shear stress

$$\sigma_{1,2} = \frac{1}{2}\left(\sigma_\varphi + \sigma_z\right) \pm \frac{1}{2}\sqrt{\left(\sigma_\varphi - \sigma_z\right)^2 + 4\tau^2}$$

$$= \frac{3}{4}\sigma_\varphi \pm \frac{1}{4}\sigma_\varphi\sqrt{1 + \left(\frac{4\tau}{\sigma_\varphi}\right)^2}, \quad \sigma_3 = 0,$$

or in normalized expressions

$$\bar{\sigma}_{1,2} = 3 \pm \sqrt{1 + \bar{\tau}^2}, \quad (\bar{\bullet}) = 4(\bullet)/\sigma_\varphi.$$

The different equivalent stresses are

(a) $\sigma_{eq} = \sigma_1 \quad \rightarrow \quad \bar{\sigma}_{eq} = 3 + \sqrt{1 + \bar{\tau}^2}.$

(b) $\sigma_{eq} = \sigma_1 - \sigma_3 \quad \rightarrow \quad \bar{\sigma}_{eq} = \begin{cases} 3 + \sqrt{1 + \bar{\tau}^2}, & 0 \leqslant \bar{\tau} \leqslant \bar{\tau}_0 \\ 2\sqrt{1 + \bar{\tau}^2}, & \bar{\tau}_0 \leqslant \bar{\tau} \end{cases}$

(c) $\sigma_{eq} = \sqrt{\sigma_1^2 + \sigma_2^2 - 2\nu\sigma_1\sigma_2}, \quad \nu = \frac{1}{3} \quad \rightarrow \quad \bar{\sigma}_{eq} = \frac{2}{\sqrt{3}}\sqrt{11 + \bar{\tau}^2},$

(d) $\sigma_{eq} = \sqrt{\sigma_1^2 - \sigma_1\sigma_2 + \sigma_2^2}, \quad \rightarrow \quad \bar{\sigma}_{eq} = \sqrt{12 + 3\bar{\tau}^2},$

If in uniaxial case yielding of the material commences at σ_0, and if we wish to avoid this situation, the different relations for the equivalent stresses define different bounds for multiaxial loading

$$\sigma_{eq} \leq \sigma_0.$$

(a)　$\dfrac{3}{4}\sigma_\varphi + \dfrac{1}{4}\sqrt{\sigma_\varphi^2 + 16\,\tau^2} \leq \sigma_0$　$\rightarrow$　$\dfrac{3}{4}\tilde{\sigma}_\varphi + \dfrac{1}{4}\sqrt{\tilde{\sigma}_\varphi^2 + 4\,\tilde{\tau}^2} \leq 1$,

(a)　$\dfrac{1}{2}\sqrt{\sigma_\varphi^2 + 16\,\tau^2} \leq \sigma_0$　$\rightarrow$　$\dfrac{1}{2}\sqrt{\tilde{\sigma}_\varphi^2 + 4\,\tilde{\tau}^2} \leq 1$,

(a)　$\dfrac{1}{2\sqrt{3}}\sqrt{11\sigma_\varphi^2 + 16\,\tau^2} \leq \sigma_0$　$\rightarrow$　$\dfrac{1}{2\sqrt{3}}\sqrt{11\tilde{\sigma}_\varphi^2 + 4\,\tilde{\tau}^2} \leq 1$,

(a)　$\dfrac{\sqrt{3}}{2}\sqrt{\sigma_\varphi^2 + 4\,\tau^2} \leq \sigma_0$　$\rightarrow$　$\dfrac{\sqrt{3}}{2}\sqrt{\tilde{\sigma}_\varphi^2 + \tilde{\tau}^2} \leq 1$,

where

$$\tilde{\sigma}_\varphi = \frac{\sigma_\varphi}{\sigma_0}, \quad \tilde{\tau} = \frac{2\tau}{\sigma_0}.$$

These bounds are depicted with Fig. 2.4.

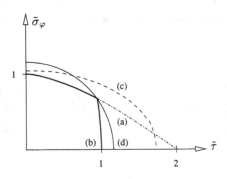

Fig. 2.4
Load-interaction curves for different strength criteria

　　It turns out that the Tresca criterion gives the lowest (most conservative) bounds to loading, and the v. Mises criterion shows similar values. For small shear stress (torsion) up to $\tau/\sigma_0 = \sqrt{2}/3$, the Tresca criterion and the maximum principal stress criterion are identical. In this regime, the four criteria produce almost similar results, whereas for larger values of the shear stress the maximum principal stress as well as the total strain energy criteria deviate considerably from the Tresca and the v. Mises criteria.

2.7 Principle of Complementary Virtual Work

In Section 1.7, we introduced the principle of virtual work as an alternative concept to solve structural problems. We emphasize again that this result is independent of any constraint of constitutive relations.

In the linear theory of elasticity, a second principle can be formulated, which is a dual counterpart of the principle of virtual work. We refer to the (Legendre-) transformation between the complementary energy and the strain energy (Eq. 2.22)

$$w^*(\sigma_{ik}) = \frac{1}{\rho} \sigma_{ik}\varepsilon_{ik} - w(\varepsilon_{ik}), \tag{2.49}$$

and, instead of virtual displacements (and strains), we now deal with virtual stresses $\delta\sigma_{ik}$ which as before are imaginary (infinitesimal) quantities which here, however, contrary to the latter case satisfy the equilibrium conditions in the interior of the body, and the dynamical boundary conditions.

Since this principle is a dual counterpart of the principle of virtual work, we start by multiplying the kinematic relations by $\delta\sigma_{ik}$ and thus arrive at

$$\int_V \left\{ \varepsilon_{ik} - \frac{1}{2}\left(\frac{\partial u_k}{\partial x_i} + \frac{\partial u_i}{\partial x_k} \right) \right\} \delta\sigma_{ik} \, dV = 0. \tag{2.50}$$

The first term of this equation gives

$$\int_V \varepsilon_{ik}\delta\sigma_{ik} \, dV = \delta W^*, \tag{2.51}$$

the complementary virtual work of the internal forces.

Due to the symmetry of the stresses, the second term is

$$\int_V \frac{\partial u_k}{\partial x_i} \delta\sigma_{ik} \, dV = \int_V \frac{\partial}{\partial x_i}(u_k\delta\sigma_{ik}) \, dV - \int_V u_k \frac{\partial}{\partial x_i}(\delta\sigma_{ik}) \, dV \tag{2.52}$$

$$= \int_A u_k \, \delta s_k \, dA - \int_V u_k \frac{\partial}{\partial x_i}(\delta\sigma_{ik}) \, dV,$$

applying the chain rule, and the divergence theorem. The virtual stresses satisfy the equilibrium conditions. Thus we find

$$\int_A u_k \, \delta s_k \, dA + \int_V u_k\rho \, \delta b_k \, dV = \delta A_A^* + \delta A_V^*, \tag{2.53}$$

the complementary virtual work of the external and the body forces, and finally

$$\delta W^* - \delta A^* = \delta(W^* - A^*) = 0. \tag{2.54}$$

This principle may also be stated alternatively in the following manner: If $\delta(W^* - A^*)$ vanishes for any arbitrary infinitesimal virtual stresses satisfying the prescribed dynamical constraints (equilibrium conditions and boundary conditions), the mechanical system fulfills the geometrical conditions (kinematical relations and boundary conditions). It is worthy of special mention that this principle of complementary virtual work holds irrespective of the stress-strain relations.

For linear elastic material, however, we find

$$w(\varepsilon_{ik}) = w^*(\sigma_{ik}) \quad \rightarrow \quad W = W^*, \tag{2.55}$$

and thus using Hooke's law, Eqs. (2.18)-(2.21), it is obvious that the strain energy density $w(\varepsilon_{ik})$ and the complementary energy density $w^*(\sigma_{ik})$ are positive definite. Thus the principle of complementary virtual work can be transformed into

$$\delta \int_V \rho w^*(\sigma_{ik}) \, dV - \int_A u_k \, \delta s_k \, dA = 0, \tag{2.56}$$

when the body forces b_k are kept fixed during a variation of stresses. Furthermore, since the u_k are kept unchanged during variation, we find a further variational principle

$$\boxed{\delta \Pi^* = 0,} \tag{2.57}$$

the principle of minimum complementary energy, with

$$\Pi^* = \int_V \rho w^*(\sigma_{ik}) \, dV - \int_A u_k s_k \, dA. \tag{2.58}$$

Among all sets of admissible stresses σ_{ik} satisfying the equilibrium conditions and the prescribed dynamical boundary conditions, the set of actual stresses makes the total complementary energy Π^* an absolute minimum.

This principle, which also is called the principle of Castigliano & Menabrea, can now be used e.g. to derive a powerful tool for solving statically indeterminate structural problems. From Eq. (2.56), we find

$$\delta W^* = \delta \int_V \rho w^* \, dV = \sum_i u_i \, \delta F_i, \quad (i, 1, 2, \ldots, n) \tag{2.59}$$

if the stresses acting on the surface of the body are reduced to a number of (generalized) forces F_i.

$$W^* = W^*(F_i) \quad \rightarrow \quad \delta W^* = \frac{\partial W^*(F_i)}{\partial F_k} \delta F_k. \tag{2.60}$$

Comparing both results, we find

$$\boxed{u_k = \frac{\partial W^*(F_i)}{\partial F_k},} \tag{2.61}$$

Engesser's theorem. A similar theorem, where the complementary energy has been expressed through the strain energy, has been formulated by Castigliano (Castigliano's first theorem). Finally, we also mention Castigliano's second theorem

$$\boxed{F_k = \frac{\partial W(u_i)}{\partial u_k},} \tag{2.62}$$

which again is the dual counterpart of Engesser's theorem, and may be derived in the same manner from the principle of virtual work.

We emphasize here that in deriving the above theorems the forces F_i, and the displacements u_i as well, have been introduced as generalized dual quantities, i.e. concentrated moments M_i, and rotations φ_i, respectively, are included with this notation.

Castigliano's second theorem (2.62) provides a method for utilizing strain energy even in the analysis of nonlinear structures. The method is based upon the use of joint displacements as the unknown quantities, which is consistent with the fact that the strain energy W must be expressed as a function of displacements u_i.

In order to develop the method, let us assume that we have a structure with n unknown joint displacements $u_1, u_2, \ldots u_n$. Assume also that the loads on the structure, denoted $F_1, F_2, \ldots F_n$, are loads that correspond to these kinematic unknowns. Then it is possible to express the strain energy W of the structure in terms of the unknown joint displacements. In the special case of a linear structure this will be a quadratic function of the displacements, as explained in Section 2.2.

Applying now Castigliano's theorem with respect to each displacement, we obtain a set of n simultaneous equations (2.62) which actually represent equilibrium conditions for each of the different forces F_k $(k = 1, 2, \ldots, n)$. The final step in the analysis is to calculate the dual quantities, such as reactions and stress resultants, from the joint displacements. It is, therefore, the displacement method of analysis, which will be illustrated in the following example.

Example 2.4:

For the truss given in the figure, using Castigliano's theorem, compute the vertical, and horizontal displacements of point A. Assume linear elastic material. Each bar has length l and cross-sectional area A, and angle $\alpha = 45°$.

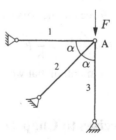

Solution:

The truss has only two degrees of freedom for joint translation, namely the horizontal and vertical translations u_1 and u_2 at joint A. In order to express the strain energy W as function of u_1 and u_2, we assume that u_1 occurs alone. Under these conditions the elongations of the bars are as follows

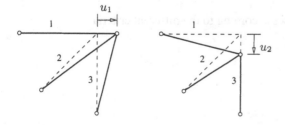

$$\Delta l_1 = u_1, \quad \Delta l_2 = \frac{1}{2}\sqrt{2}\,u_1, \quad \Delta l_3 = 0,$$

under the assumption of small deformations. When the displacement u_2 occurs alone, the elongations are

$$\Delta l_1 = 0, \quad \Delta l_2 = -\frac{1}{2}\sqrt{2}\,u_2, \quad \Delta l_3 = -u_2.$$

When both u_1 and u_2 occur simultaneously, the elongations are

$$\Delta l_1 = u_1, \quad \Delta l_2 = \frac{1}{2}\sqrt{2}\,(u_1 - u_2), \quad \Delta l_3 = -u_2.$$

Thus from Eq. (2.19) (see also Chapter 6), the strain energy $W(u_k)$ can be obtained by summing the energies for the three bars

$$W = \frac{EA}{2l}\left\{u_1^2 + \frac{1}{2}(u_1 - u_2)^2 + u_2^2\right\} = \frac{EA}{2l}\left\{\frac{3}{2}u_1^2 - u_1 u_2 + \frac{3}{2}u_2^2\right\}.$$

Applying now Castigliano's theorem, we find

$$F_1 = \frac{EA}{2l}\left\{3u_1 - u_2\right\} = 0$$

$$F_2 = \frac{EA}{2l}\left\{3u_2 - u_1\right\} = F,$$

and from these equations

$$u_1 = \frac{1}{8}\frac{Fl}{EA}, \quad u_2 = \frac{3}{8}\frac{Fl}{EA}.$$

Finally, the forces in the bars can be calculated to be

$$F_1 = EA\frac{\Delta l_1}{l} = \frac{1}{8}F, \quad F_2 = -\frac{1}{8}\sqrt{2}F, \quad F_3 = -\frac{3}{8}F.$$

In passing we mention that we just have solved a statically indeterminate structure.

2.8 Exercises to Chapter 2

Problem 2.1:

The stress tensor at a point in plane stress is

$$\sigma = \begin{pmatrix} \sigma & \tau \\ \tau & 0 \end{pmatrix}.$$

Determine the equivalent stresses according to the different criteria

1. maximum principal stress,
2. maximum shear stress,
3. total strain energy, and
4. shear strain energy criterion.

Problem 2.2:

Compute the strain energy function W of the truss given in the figure. All bars of the truss have the same cross-sectional area A and Young's modulus E.

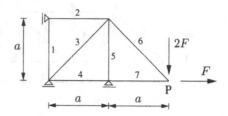

Problem 2.3:

Compute - using Castigliano's theorem - the vertical, and horizontal displacements for point P of the truss of Problem 2.2. Assume linear elastic material behaviour.

Problem 2.4:

After the truss in the figure is mounted, there is an increase in temperature of θ_0. The bar 1 has a length of a and all the other bars are of length $2a$. All bars have the same cross-sectional area A, Young's modulus E, and linear heat expansion coefficient α. Compute using Castigliano's theorem the vertical, and horizontal displacements of point P. Assume linear elastic material behaviour.

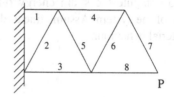

Problem 2.5:

For an increase in temperature of Θ, compute the vertical, and horizontal displacements of point P. All bars of the truss have the same cross-sectional area A, Young's modulus E, and linear heat expansion coefficient α. Assume linear elastic material behaviour.

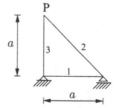

Problem 2.6:

Compute the strain energy function W of the truss given below. All bars of the truss have the same cross-sectional area A and Young's modulus E. Compute – using Castigliano's theorem – the vertical, and horizontal displacements of point P. Assume linear elastic material behaviour.

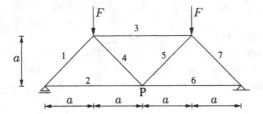

Problem 2.7:

The Point P is supported in a statically indeterminate manner by the bars 1, 2, and 3 as shown in the figure. All bars have the same cross-sectional area A and Young's modulus E. Compute the vertical, and horizontal displacements of point P. Also compute the strain energy function W of the system. Assume linear elastic material behaviour.

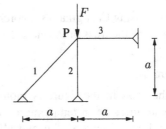

3. The Theory of Simple Beams I

3.1 General, Normal Stresses

A simple beam is defined as a single component of a structure, in which one dimension (the axis of the beam) is long compared with the other two dimensions. In this chapter, we shall limit our scope to long straight beams, and - for the moment - we will assume that the xz-plane is a plane of symmetry of the beam and that the loading acts in this same plane; hence the bending deflections will take place in this plane as well.

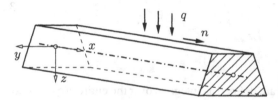

Fig. 3.1
Beam of constant cross section

At first, beams of constant cross section (prismatic beams) will be considered, the center of the cross section area being coincident with the x-axis.

The assumptions of the simple theory of beams are:

(i)   the beam is prismatical and straight with symmetry of the cross section with respect to the xz-plane,

(ii) the y and z axes are principal axes of the cross section,

(iii) the stress resultants are

$$N = \text{const.} \qquad M_x = 0$$
$$Q_z = Q = \text{const.} \qquad Q_y = 0$$
$$M_y = M = M(x) \qquad M_z = 0.$$

For pure bending, i.e. $M = \text{const.}$, $Q = 0$, $N = 0$, we know that for symmetry reasons the axis of the beam is bent into a circular arc, and any cross section remains plane and normal to the longitudinal fibers of the beam. Thus we further assume - even for those cases, where we have shear forces and axial forces acting on the cross section:

(iv) plane cross sections normal to the x axis in the unstressed state remain plane and normal to the deformed beam axis (Bernoulli's hypothesis),

(v) intersecting planes parallel to the beam axis remain stress-free.

From these assumptions, we can conclude (assumption (iv))

$$\varepsilon_{xy} = \varepsilon_{xz} = 0 \tag{3.1}$$

and hence (Hooke's law)

$$\sigma_{xy} = \sigma_{xz} = 0, \tag{3.2}$$

and further from assumption (v)

$$\sigma_{yy} = \sigma_{zz} = \sigma_{yz} = 0, \tag{3.3}$$

i.e., there is only one remaining component of the stress tensor, namely $\sigma_{xx} \neq 0$.

Introducing the radius of curvature R of any element along the axis of the beam, and further assuming that the deflections and slopes are small so that the equation for the curvature can be approximated to

$$\frac{1}{R} \cong -w''(x) \tag{3.4}$$

where w is the displacement of the beam axis in the z direction, and the prime designates differentiation of a function with respect to x, we arrive at

$$\varepsilon_{xx} = \varepsilon_0(x) + \frac{z}{R} \tag{3.5}$$

and hence, since we have $\sigma_{xx} = E\varepsilon_{xx}$

$$\sigma_{xx} = E\left[\varepsilon_0(x) + \frac{z}{R}\right]. \tag{3.6}$$

The stress resultants N and M are defined as integrals over the entire cross-sectional area

$$N = \int_A \sigma_{xx}\, dA, \quad M_y = \int_A z\sigma_{xx}\, dA. \tag{3.7}$$

Thus we find

$$N = \int_A \sigma_{xx}\, dA = E\varepsilon_0(x) \int_A dA + \frac{E}{R} \int_A z\, dA, \tag{3.8}$$

$$M_y = \int_A z\sigma_{xx}\, dA = E\varepsilon_0(x) \int_A z\, dA + \frac{E}{R} \int_A z^2\, dA, \tag{3.9}$$

in which

$$J_{yy} = J = \int_A z^2\, dA, \quad A = \int_A dA \tag{3.10}$$

are the moment of inertia with respect to the y axis and the definition of the area, respectively. We further emphasize that the first moment of the area of the cross section with respect to the y axis

$$S_y = \int_A z\, dA = 0 \tag{3.11}$$

vanishes for a system of centroidal axes. Thus from above

$$N = EA\varepsilon_0(x) = EAu'(x), \quad M = \frac{EJ}{R}, \tag{3.12}$$

where u is the displacement of the beam axis in the x direction, and finally from Eq. (3.6)

$$\sigma_{xx} = \sigma_{xx}(x, z) = \frac{N}{A} + \frac{M(x)}{J} z, \tag{3.13}$$

describing the relation between the normal stresses in the beam and the stress resultants. Introducing the second part of Eq. (3.12) into Eq. (3.4), we find

$$M(x) = -EJ w''(x). \tag{3.14}$$

Equations $(3.12)_1$ and (3.14) constitute the basic differential equations for the deflections u and w of the beam - more precisely of the beam axis - in the xz plane.

This result can be generalized to include bending with respect to the z axis

$$M_z = -\int_A y\sigma_{xx} \, dA \tag{3.15}$$

to give

$$\sigma_{xx} = \sigma_{xx}(x, z) = \frac{N}{A} + \frac{M_y(x)}{J_{yy}} z - \frac{M_z(x)}{J_{zz}} y, \tag{3.16}$$

provided both x and y axes are principal axes, for which the product of inertia

$$J_{yz} = J_{zy} = -\int_A zy \, dA = 0 \tag{3.17}$$

vanishes.

From the condition $\sigma_{xx} = 0$, the neutral axis can be determined

$$0 = \frac{N}{A} + \frac{M_y(x)}{J_{yy}} z - \frac{M_z(x)}{J_{zz}} y, \tag{3.18}$$

which is a straight line passing through the centroid of the cross section if $N = 0$.

In the preceding analysis according to assumption (iv), i.e. Eqs. (3.1) and (3.2), the shear stresses σ_{xz} as well as σ_{xy} and thus their resultant shear forces Q_z and Q_y, which are defined as the following integrals

$$Q_z = \int_A \sigma_{xz} \, dA, \quad Q_y = \int_A \sigma_{xy} \, dA \tag{3.19}$$

have been ignored.

Deformations associated with these shear stresses would consist of a warping of the cross section so that the cross section that was plane before bending no longer remains plane during bending. This warping complicates the description of the behaviour, but more elaborate analyses show that the normal stresses calculated from the flexure formula (Eqs. 3.13 and 3.16, respectively) are not significantly altered by the presence of the shear stresses. Thus, it is justifiable to use Bernoulli's hypothesis even when we have non-uniform bending.

We note, however, that this generally accepted beam theory with regard to the shear stresses is inconsistent with Hooke's law. According to Bernoulli's hypothesis shear strains have to vanish. The shear forces acting on the cross section - on the other hand - imply shear stresses, which thus have to be introduced independently.

In the above calculations, we have made use of the definitions of the moments of inertia, which are introduced as second moments of the (plane) area of the cross section

$$J_{yy} = \int_A z^2 \, \mathrm{d}A, \quad J_{zz} = \int_A y^2 \, \mathrm{d}A, \tag{3.20}$$

with respect to the y and the z axes, respectively, which from now on are no longer assumed as principal axes, and the product of inertia [1]

$$J_{yz} = J_{zy} = -\int_A yz \, \mathrm{d}A, \tag{3.21}$$

with respect to the same centroidal axes.

In passing, we realize that these quantities can also be interpreted as components of a two dimensional second-rank tensor through

$$J_{ik} = \int_A \{(x_r x_r)\delta_{ik} - x_i x_k\} \, \mathrm{d}A = \begin{pmatrix} J_{yy} & J_{yz} \\ J_{zy} & J_{zz} \end{pmatrix}, \quad x_i = y, z, \tag{3.22}$$

and thus follows the same transformation rules, and has similar properties as any (two dimensional) second-rank tensor, e. g. the stress tensor.

From similar relations, introduced with Eqs. (1.16) and (1.17) for a plane stress state, we can determine the directions of the principal axes with the extreme values called the principal values of inertia. It is also found that the product of inertia vanishes in these principal directions. In the preceding calculations, this fact has already been used.

The moment of inertia about the centroid C is called the polar moment of inertia with respect to the origin

$$J_0 = J_{yy} + J_{zz} = \int_A (y^2 + z^2) \, \mathrm{d}A = \int_A r^2 \, \mathrm{d}A. \tag{3.23}$$

It is observed that J_0 is independent from any rotation and thus an invariant of J_{ik}.

Moreover, there exists a relation between the moments of inertia of a plane area about a centroidal axis and the moments of inertia of the same area about an axis parallel to the centroidal axis (Fig. 3.2). This relation is expressed by the equations

---

[1] It should be noted that the product of inertia is defined here with a negative sign before the integral, whereas, in many other textbooks and tables, it is defined without the minus sign. The reason is that here and according to Eq. (3.22) the moments of inertia are interpreted as components of a second-rank tensor.

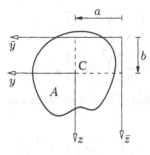

Fig. 3.2
Cross section with parallel axes

$$J_{\bar{y}\bar{y}} = J_{yy} + b^2 A$$
$$J_{\bar{z}\bar{z}} = J_{zz} + a^2 A \tag{3.24}$$
$$J_{\bar{y}\bar{z}} = J_{yz} - abA \,,$$

where J_{yy}, J_{zz} are the moments of inertia of the area A for the y and z axes through the centroid C, and J_{yz} is the product of inertia for the same axes. Equation (3.24) is known as the parallel-axis theorem.

Example 3.1:

For the cross section shown in the figure, compute the principal moments of inertia and planes:

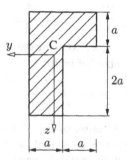

Solution:

We first divide the cross section into two rectangles of dimension $a \times 2a$, and determine the coordinates of the common centroid with the help of relations

$$\bar{z}_C = \frac{\sum_i \bar{z}_{Ci} A_i}{\sum_i A_i} = \frac{a/2 \cdot 2a^2 + 2a \cdot 2a^2}{4a^2} = \frac{5}{4} a$$

$$\bar{y}_C = \frac{\sum_i \bar{y}_{Ci} A_i}{\sum_i A_i} = \frac{a \cdot 2a^2 + 3a/2 \cdot 2a^2}{4a^2} = \frac{5}{4} a \,,$$

where the $\bar{y}$ and $\bar{z}$ coordinates, respectively, are the distances measured with respect to a parallel system with its origin in the upper right corner of the cross section. Now using the parallel-axis theorem (Eq. 3.24), we can determine the moments of inertia with respect to the y and z axes, respectively, as the sum of the moments of the two separate rectangles

$$J_{yy} = \frac{a(2a)^3}{12} + 2a^2\left(\frac{3}{4}a\right)^2 + \frac{2a\,a^3}{12} + 2a^2\left(\frac{3}{4}a\right)^2 = \frac{37}{12}a^4$$

$$J_{zz} = \frac{2a\,a^3}{12} + 2a^2\left(\frac{1}{4}a\right)^2 + \frac{a(2a)^3}{12} + 2a^2\left(\frac{1}{4}a\right)^2 = \frac{13}{12}a^4.$$

Due to the symmetry of the two rectangles their product of inertia vanishes and we thus get from the parallel-axis theorem (Eq. 3.24₃)

$$J_{yz} = 0 - 2a^2\left(\frac{1}{4}a\right)\left(\frac{3}{4}a\right) + 0 - 2a^2\left(-\frac{3}{4}a\right)\left(-\frac{1}{4}a\right) = -\frac{3}{4}a^4.$$

The principal values of inertia can be computed by applying Eq. (1.17) to the moments of inertia. We thus arrive at

$$J_{1,2} = \frac{1}{2}(J_{yy} + J_{zz}) \pm \frac{1}{2}\sqrt{(J_{yy} - J_{zz})^2 + 4J_{yz}^2}$$

$$= \frac{25}{12}a^4 \pm \frac{1}{2}\sqrt{4a^8 + \frac{9}{4}a^8} = \frac{a^4}{12}(25 \pm 15),$$

and finally, since $J_1 \geqslant J_2$,

$$J_1 = \frac{10}{3}a^4, \quad J_2 = \frac{5}{6}a^4.$$

Likewise, we compute the directions of the principal axes by applying Eq. (1.16) to the moments of inertia

$$2\varphi = \arctan\frac{2J_{yz}}{J_{yy} - J_{zz}} = \arctan\left(-\frac{3}{4}\right).$$

Since the solution of Eq. (1.16) is not unique, Table 1.1 helps us in finding the magnitude of the inclination angle. We have

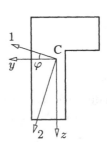

$$J_{yy} \geqslant J_{zz} \quad \text{and} \quad J_{yz} \leqslant 0 \quad \rightarrow \quad -\frac{\pi}{4} \leqslant \varphi \leqslant 0,$$

and thus we find

$$2\varphi = -36.9° \quad \rightarrow \quad \varphi = -18.44°.$$

$\square$

In passing, we emphasize that the well-known relations for the normal stresses Eqs. (3.13) and (3.16) as well as the differential equation (3.14) describing the deflections of the beam axis are only valid within systems of principal axes. This means that as in the preceding example, if the geometry and the loading are described in non-principal centroidal axes, these quantities - including the loading - have to be transformed to principal axes, prior to applying Eqs. (3.13) and (3.16), and (3.14), respectively.

Example 3.2:

We will demonstrate this with a bending moment of $M_y = 20\,\text{kNm}$ applied to the cross section of Example 3.1, with dimension $a = 20\,\text{cm}$.

Solution:

Since the y and z axes are no principal axes, the given bending moment is to be resolved into components of the 1 $(\bar{y})$ and 2 $(\bar{z})$ axes

$$M_{\bar{y}} = \quad M_y \cos\varphi = 18.974\,,$$
$$M_{\bar{z}} = -M_y \sin\varphi = \quad 6.325\,.$$

Introducing these values into Eq. (3.16) yields

$$\sigma_{xx} = \frac{M_{\bar{y}}}{J_1}\,\bar{z} - \frac{M_{\bar{z}}}{J_2}\,\bar{y} = 3.558\,\bar{z} - 4.743\,\bar{y}\,.$$

The neutral axis of this stress distribution is described with ($\sigma_{xx} = 0$)

$$\bar{z} = 1.333\,\bar{y}\,.$$

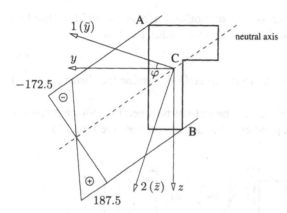

Fig. 3.3
Distribution of normal stress, and neutral axis

The extreme values of these stresses will be found at points A and B of the cross section, where the transformed coordinates may be calculated according to the transformation rule

$$\bar{y} = z\sin\varphi + y\cos\varphi$$
$$\bar{z} = z\cos\varphi - y\sin\varphi\,.$$

We thus find

$$\text{A}: \quad (1.107\,a\,;\ -0.949\,a) \quad \rightarrow \quad \sigma_{xx} = -172.50\,\text{N/cm}^2,$$
$$\text{B}: \quad (-0.791\,a\,;\ 1.581\,a) \quad \rightarrow \quad \sigma_{xx} = \quad 187.50\,\text{N/cm}^2.$$

A 'naive' calculation with the stresses simply calculated from

$$\sigma_{xx} = \frac{M_y}{J_{yy}}\,z = \begin{cases} -101.4\,\text{N/cm}^2 \\ 141.9\,\text{N/cm}^2 \end{cases}$$

would underestimate these extreme values considerably.

3.2 Shear Stresses

In the preceding section, we determined the normal stresses of the beam. We now will investigate the distribution of the shear stresses, and begin with the simplest case of a beam of rectangular cross section having width b and height h. It is natural to assume that for this beam the shear stresses are parallel to the shear force Q, i.e. parallel to the vertical sides of the cross section. As a second assumption, we take

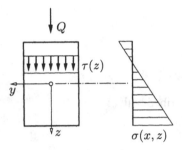

Fig. 3.4
Normal and shear stresses of a beam with rectangular cross section

the distribution of the shear stresses to be uniform across the width of the beam. Thus – inconsistent with Hooke's law – we replace Eq. (3.2) by

$$\sigma_{xy} = 0, \quad \sigma_{xz} = \tau(z). \tag{3.25}$$

These assumptions will enable us to completely determine the distribution of the shear stresses.

A small element of length dx may be cut out between two adjacent cross sections and between two planes parallel to the neutral surface (see Fig. 3.5).

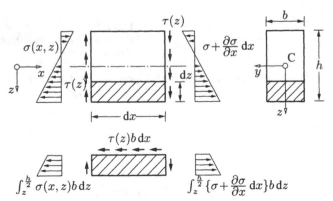

Fig. 3.5 Small element of a beam with distribution of stresses

Integrating over the relevant surfaces, the equilibrium of the forces in x direction yields

$$-\int_{z}^{\frac{h}{2}} \sigma(x,z)b\,dz + \int_{z}^{\frac{h}{2}} \left\{ \sigma(x,z) + \frac{\partial\sigma}{\partial x}\,dx \right\} b\,dz - \tau(z)b\,dx = 0,$$

from which

$$\tau(z) = \frac{1}{b} \int_z^{\frac{h}{2}} \frac{\partial \sigma}{\partial x} b \, dz = \frac{1}{b} \int_z^{\frac{h}{2}} \frac{\partial \sigma}{\partial x} \, dA. \tag{3.26}$$

The partial differential of σ can be determined from Eq. (3.13)

$$\frac{\partial \sigma}{\partial x} = \frac{\partial}{\partial x} \left\{ \frac{N}{A} + \frac{M(x)}{J} z \right\} = \frac{dM}{dx} \frac{z}{J} = \frac{Q}{J} z, \tag{3.27}$$

since N as well as the width and height of the rectangular cross section are assumed to be constant. Introducing this relation into Eq. (3.26), we find

$$\tau(z) = \frac{Q}{Jb} \int_z^{\frac{h}{2}} z \, dA. \tag{3.28}$$

The integral in this equation represents the first moment of the shaded portion of the cross section with respect to the y axis, i.e. the first moment of the cross-sectional area below the arbitrary level z

$$S_y(z) = \int_z^{\frac{h}{2}} z \, dA. \tag{3.29}$$

With this definition, we can write

$$\boxed{\tau(z) = \frac{Q S_y(z)}{Jb}.} \tag{3.30}$$

For the rectangular cross section under consideration, the quantity S_y is

$$S_y(z) = \int_z^{\frac{h}{2}} bz \, dz = \frac{bh^2}{8} \left\{ 1 - \left(\frac{2z}{h} \right)^2 \right\}, \quad J = \frac{bh^3}{12}. \tag{3.31}$$

Now substituting this expression into Eq. (3.30), we get

$$\boxed{\tau(z) = \frac{3}{2} \frac{Q}{A} \left\{ 1 - \left(\frac{2z}{h} \right)^2 \right\}.} \tag{3.32}$$

This result shows that the shear stress varies parabolically with z. The stress is zero when $z = \pm h/2$ and has its maximum value

$$\tau_{max} = \frac{3}{2} \frac{Q}{A} \tag{3.33}$$

at $z = 0$.

For more general, but still symmetric cross sections there is no longer any basis for assuming that all of the shear stresses are parallel to the z axis. In fact, we can easily show that at a point on the boundary of the cross section, the shear stress must be tangent to the boundary, since the outer surface of the beam is free from any stresses if there are no distributed forces acting on these surfaces.

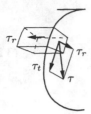

Fig. 3.6
Boundary conditions

We therefore have to modify assumption (3.25) in such a way that the symmetry of the stress distributions is guaranteed, and, further, that the resultant τ at the boundary of the cross section is always tangent to this boundary

$$\sigma_{xz} = \sigma_{xz}(z), \quad \sigma_{xy} = \sigma_{xz}\frac{db(z)}{dz}\frac{y}{b(z)}. \tag{3.34}$$

Since the first part of assumption (3.34) coincides completely with that made for the rectangular cross section, we can use Eq. (3.30) in calculating this component, i.e.

$$\boxed{\sigma_{xz}(z) = \frac{QS_y(z)}{b(z)J}} \tag{3.35}$$

and

$$\boxed{\sigma_{xy}(y,z) = \sigma_{xz}\frac{db(z)}{dz}\frac{y}{b(z)}.} \tag{3.36}$$

Example 3.3:

Let us demonstrate the validity of our assumptions for a prismatic beam with a symmetric triangular cross section (with width B and height H).

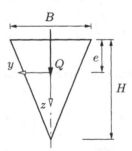

Solution:

The distance e from the base to the centroid of the cross section is

$$e = \frac{H}{3}$$

and the moment of inertia J of the cross section about the y axis is

$$J = J_{yy} = \frac{BH^3}{36}.$$

With the width

$$b(z) = \frac{2}{3}B - \frac{B}{H}z,$$

we determine the first moment of the cross-sectional area (Eq. 3.29)

$$S_y(z) = \int_z^{\frac{3}{2}H} b(z)z\,\mathrm{d}z = BH^2\left\{\frac{4}{81} - \frac{1}{3}\left[\left(\frac{z}{H}\right)^2 - \left(\frac{z}{H}\right)^3\right]\right\}.$$

Therefore, the shear stresses σ_{xz} in the triangular cross section are (Eq. 3.35)

$$\sigma_{xz}(z) = \frac{4}{3}\frac{Q}{A}\left(1 - \frac{3}{2}\frac{z}{H}\right)\left(1 + 3\frac{z}{H}\right).$$

The shear stress σ_{xz} varies parabolically with z. The stress is zero when $z = -H/3$ and $z = 2H/3$, respectively. The maximum is obtained at $z = H/6$

$$\sigma_{xz\,\mathrm{max}} = \frac{3}{2}\frac{Q}{A}.$$

To determine the second component σ_{xy}, we calculate

$$\frac{\mathrm{d}b(z)}{\mathrm{d}z} = -\frac{B}{H}.$$

Now, from Eq. (3.36), we obtain

$$\sigma_{xy}(y,z) = -2\frac{Q}{A}\left(1 + 3\frac{z}{H}\right)\frac{y}{H}.$$

This result varies linearly with y and z, and vanishes for $y = 0$ and $z = -H/3$.
    At the right boundary, e.g. we have

$$y = -\frac{1}{2}b(z) \longrightarrow \sigma_{xy} = \sigma_{xz}\frac{B}{2H}.$$

Thus the resultant shear stress τ is determined from

$$\tau = \sqrt{\sigma_{xz}^2 + \sigma_{xy}^2} = \sigma_{xz}\sqrt{1 + \left(\frac{B}{2H}\right)^2}$$

and the direction of τ runs with

$$\tan\alpha = \frac{\sigma_{xy}}{\sigma_{xz}} = \frac{B}{2H}$$

parallel to this boundary.

Example 3.4:
For a circular cross section with radius R and area $A = \pi R^2$, we find

$$\sigma_{xz} = \frac{4}{3}\frac{Q}{A}\left\{1 - \left(\frac{z}{R}\right)^2\right\}$$

$$\sigma_{xy} = -\frac{4}{3}\frac{Q}{A}\frac{yz}{R^2},$$

with the maximum value of the resultant shear stress τ

$$\tau_{\mathrm{max}} = \frac{4}{3}\frac{Q}{A}$$

on the y axis (at $z = 0$).

$\square$

Finally, we will consider a particular class of beams, known as beams of thin-walled cross sections, for which we can determine the shear stresses by the same method used in deriving Eq. (3.30). The beams to be considered are distinguished by two criteria:

(i) the cross section has a thickness δ which is small compared to its overall height or width, and

(ii) the cross section is either open (a) or closed (b)

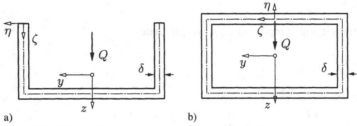

Fig. 3.7 Thin-walled cross sections: (a) open, (b) closed

Beams of thin-walled cross sections are widely used in engineering applications.

In both cases, the y and z axes are centroidal principal axes of the cross sections, and the load Q acts in the axis of symmetry. According to criterion (i), in addition to the y and z axes, we now introduce two new axes, which are suited better for a mechanical description of the profile of the cross section. These are the ζ axis, following the (dashed) cross-sectional centerline, and the η axis perpendicular to ζ.

This new (rectangular) description allows us to modify the assumption about the shear stresses in the following way

$$\sigma_{x\eta} = 0, \quad \sigma_{x\zeta} = \tau(\zeta), \tag{3.37}$$

which are the same assumptions as for a rectangular cross section. Thus, any small element of length $d\zeta$ and width $\delta(\zeta)$ of the cross-sectional area can be regarded as an element of a rectangular cross section.

We start analyzing the shear stress $\tau(\zeta)$ of thin-walled open cross sections, and hence find

$$\boxed{\tau(\zeta) = \frac{QS(\zeta)}{J\delta(\zeta)} ,} \tag{3.38}$$

wherein

$$S(\zeta) = \int_\zeta^L z(\zeta)\delta(\zeta)\,d\zeta = \int_\zeta^L z(\zeta)\,dA \tag{3.39}$$

again is the first moment of the cross-sectional area below the arbitrary level ζ.

In the same manner, we can also treat thin-walled closed cross sections. Due to the assumed symmetry with respect to the xz plane, τ must vanish on the z axis, i.e. for $y = 0$. This allows us to divide the whole cross section into parts with vanishing shear stresses at both ends.

Example 3.5:

Determine the shear stresses for the T-profile shown in the figure:

$b = h = 200$ mm, $\delta = 10$ mm,

$Q = 100/3$ kN.

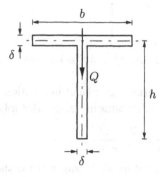

Solution:

We introduce the profile and the new axes ζ_i, $i = 1, 2, 3$, and determine the distance e to the centroid.

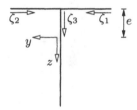

Fig. 3.8
Coordinates ζ_i

$$e = \frac{\delta h \frac{h}{2}}{2\delta h} = \frac{h}{4}$$

$$J = J_{yy} = \frac{1}{12}\,\delta h^3 + 2\,\delta h e^2 = \frac{1}{12}\,\delta h^3 + \frac{1}{8}\,\delta h^3 = \frac{5}{24}\,\delta h^3.$$

To solve this problem, we use a graphical integration method, starting with the distribution of the function $z\delta(\zeta)$ along the different ζ_i axes. Here, we have made use of the fact that the total integral of the first moment of the cross-sectional area vanishes, i.e.

$$\int_0^L z\,\delta(\zeta)\,d\zeta = \int_0^\zeta z\,\delta(\zeta)\,d\zeta + \int_\zeta^L z\,\delta(\zeta)\,d\zeta = 0.$$

Thus instead of using the definition (3.39), we have

$$S(\zeta) = -\int_0^\zeta z\,\delta(\zeta)\,d\zeta\,,$$

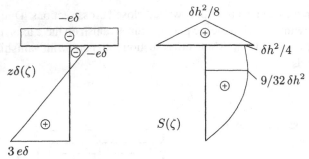

Fig. 3.9 $z\delta$, and S as functions of ζ_i

which permits graphical integration in the same way as the integration of stress resultant diagrams in statics. Multiplying this result by the factor

$$\frac{Q}{J\delta} = \frac{24}{5}\frac{Q}{\delta^2 h^3},$$

we arrive at the distribution of the shear stress τ along the different ζ_i axes, with a maximum value of $\tau = 22.5$ MPa at the centroid.

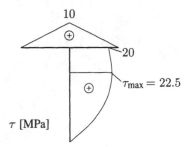

Fig. 3.10
τ as function of ζ_i

Example 3.6:

A beam with a thin-walled hollow square cross section with centerline dimensions:

$b = 500, \quad h = 900, \quad \delta = 20 \quad$ [mm]

is subject to a shear force Q. Compute the shear stresses for this profile.

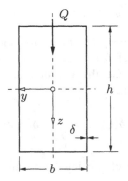

Solution:

We determine the cross-sectional area and moment of inertia to be

$$A = 2\delta(b + h) = 56 \cdot 10^3 \text{ mm}^2$$

$$J = \frac{1}{6}\delta h^3 + b\delta\frac{h^2}{2} = \frac{1}{6}\delta h^2(h + 3b) = 648 \cdot 10^7 \text{ mm}^4.$$

We first analyze the shear stress of a closed cross section. Due to the symmetry of the section with respect to the xz plane, we know that τ vanishes at $y = 0$. We therefore only need to examine one half of the section, e.g. the left half – with the shear force Q acting in z direction.

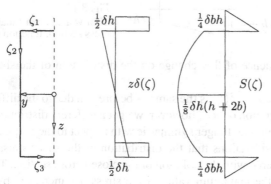

Fig. 3.11 Thin-walled square cross section, functions $z\delta$, and S

The distribution of the shear stress τ is obtained by multiplying $S(\zeta)$ by the factor

$$\frac{Q}{J\delta} = \frac{6Q}{\delta^2 h^2(h + 3b)}$$

with the maximum value of

$$\tau_{\max} = \frac{3}{4}\frac{Q}{\delta h}\frac{h + 2b}{h + 3b} = 0.594\frac{Q}{\delta h} = 1.847\frac{Q}{A}.$$

The resultant of all shear stresses on the cross section is clearly a vertical force, because the horizontal stresses in the flanges - in both parts of the section - cancel one another and produce no resultant. The shear stresses in the web have a resultant T_2, which can be found by integrating over the whole domain ζ_2, as follows

$$T_2 = \int_{\zeta_2}\tau(\zeta)\delta\,\mathrm{d}\zeta = \frac{Q}{J}\int_{\zeta_2}S(\zeta)\,\mathrm{d}\zeta = \frac{Q}{J}\frac{1}{12}\delta h^2(h + 3b) = \frac{Q}{2},$$

which establishes the fact that the resultant of the shear stresses is equal to the vertical load

$$Q = 2T_2.$$

We now assume that the square cross section is a welded structure, with a long weld in the webs at $z = 0$. We further assume that due to some carelessness during the welding process one of the welds, the right one (say), breaks, and that the closed cross section turns over to an open one. In the second part of this example, we

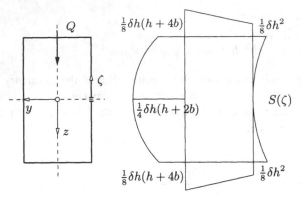

Fig. 3.12
Thin-walled open square
cross section, function S

are now interested in the influence of this change on the distribution of the shear stresses.

The distribution of the function $z\delta(\zeta)$ is the same as before, but due to the different starting points of the integration over ζ, however, we get a different distribution of the first moment $S(\zeta)$, which is no longer symmetric with respect to the xz plane.

It follows from these considerations that the distribution of the shear stress τ has changed drastically compared with the solution of the closed cross section. The symmetry has been lost and the maximum value of the stress has increased by a factor of 2.

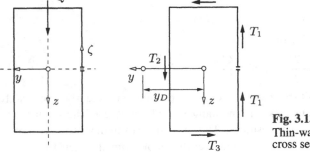

Fig. 3.13
Thin-walled open square
cross section

Moreover, if we calculate the resultants of the shear stresses in the different regimes of ζ, with

$$T_1 = \frac{Q}{J}\frac{1}{48}\delta h^3 \qquad\qquad = 0.0469\,Q$$

$$T_2 = \frac{Q}{J}\frac{1}{24}\delta h^2(5h + 12b) = 1.0938\,Q$$

$$T_3 = \frac{Q}{J}\frac{1}{8}\delta hb(h + 2b) \qquad = 0.3299\,Q,$$

we have to realize that the horizontal stresses in the flanges no longer cancel one another. Instead, in addition to the resultant of the vertical stresses

$$T = T_2 - 2T_1 = Q\,,$$

we end up with a twisting moment about the x axis, which does not exist as an external moment, thus indicating that the assumption that the beam axis is coincident with the x axis has to be modified. This inconsistency may help us to determine the shear center of a non-symmetric cross section.

3.3 Shear Center of Thin-Walled Open Sections

In the preceding section, we developed techniques and formulas for finding the shear stresses in beams of thin-walled closed and open sections. Now we will use that information to locate the shear centers of cross sections of several different shapes.

For every cross-sectional shape, there exists one specific point on the cross section through which the resultant of the transverse shear forces always passes, regardless of the direction of the transverse external forces. This point is called the shear center of the cross section, and the locus of the shear centers of the cross section of a beam is called the elastic axis of the beam. Loads acting on a beam must pass through the elastic axis in order to produce simple bending of the beam with no twisting.

It is straightforward to show that the shear center for the cross section of Example 3.6 must be located on the symmetry axis, at a distance y_D from the x axis, such that the moment of the different shear forces T_i with respect to this point vanishes (see Fig. 3.13)

$$T_2\left(y_D - \frac{b}{2}\right) - T_3 h - 2T_1\left(y_D + \frac{b}{2}\right) = 0\,.$$

From this, we may calculate

$$y_D = \frac{1}{Q}\left(T_2\frac{b}{2} + T_3 h + T_1 b\right) = \frac{3b}{2}\frac{h + 2b}{h + 3b} = 59.38 \text{ cm}.$$

The location of the shear center is not always readily obtainable. If the cross section has an axis of symmetry, the shear center will be located on this axis. If the cross section has two axes of symmetry, the shear center will coincide with the centroid.

From the above results, we may conclude

$$Q_z y_D - Q_y z_D = \int_0^L \tau(\zeta)\delta(\zeta)a(\zeta)\,\mathrm{d}\zeta \qquad (3.40)$$

for the general case of transverse loading, where $a(\zeta)$ is the distance of the specific point of the cross-sectional centerline from the elastic line of the bar, i.e. the axis of the shear centers of the cross sections. Introducing the results of Section 3.2 (Eq. (3.38) - and its generalization to multiaxial loading), we finally arrive at

$$y_D = \frac{1}{J_{yy}} \int_0^L a(\zeta) S_y(\zeta) \, d\zeta$$

$$z_D = \frac{-1}{J_{zz}} \int_0^L a(\zeta) S_z(\zeta) \, d\zeta .$$

(3.41)

If the resultant shear force does not pass through the shear center, we end up with an additional twisting moment of the magnitude resultant shear force times the distance from the shear center. In the case of Example 3.6 this would mean an additional twisting moment of

$$M_T = -Q \, y_D = -0.594 \, Q \quad [kNm].$$

Example 3.7:

For the thin-walled semicircular section shown in the figure, compute the shear stresses along the profile, and the coordinates of the shear center.

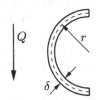

Solution:

The distribution of the shear stresses along the profile of the cross section is computed according to Eq. (3.38)

$$\tau(\zeta) = \frac{Q \, S_y(\zeta)}{J_{yy} \, \delta(\zeta)} ,$$

where here the definition (3.39) of the first moment of the cross-sectional area will be used,

$$S_y(\zeta) = \int_\zeta^L z(\zeta) \delta(\zeta) \, d\zeta .$$

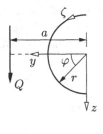

For a thin-walled structure, with $\delta = $ const., we have $dA = \delta \, d\zeta$, and thus for the moment of inertia

$$J_{yy} = \int_L z^2 \delta \, d\zeta = \int_{-\frac{\pi}{2}}^{\frac{\pi}{2}} \delta r^3 \sin^2 \varphi \, d\varphi = \frac{\pi}{2} \delta r^3,$$

where use has been made of the transformation relations $z = r \sin \varphi$ and $d\zeta = r \, d\varphi$. In the same way, we compute the first moment

$$S_y(\zeta) = \int_\zeta^L z(\zeta) \delta \, d\zeta = \delta r^2 \int_\varphi^{\frac{\pi}{2}} \sin \varphi \, d\varphi = \delta r^2 \cos \varphi .$$

From this result, the shear stresses can be determined

$$\tau(\zeta) = \frac{2Q}{\pi r \delta} \cos \varphi .$$

The distribution along the profile is shown in the diagram, where also the direction of the shear flow $t(\zeta) = \tau(\zeta)\delta$ has been sketched.

The resultant of the vertical components of this shear flow turns out to be

$$T = \int_L t(\zeta) \cos \varphi \, d\zeta = \frac{2Q}{\pi} \int_\varphi \cos^2 \varphi \, d\varphi = Q ,$$

which can also be used to control our calculations.

The horizontal components cancel each other. Obviously, however, this shear flow has a resultant twisting moment about the x-axis, which must be equivalent to the moment produced by the transverse shear force Q

$$Q \, a = \int_L t(\zeta) \, r \, d\zeta = \frac{2Q \, r}{\pi} \int_\varphi \cos \varphi \, d\varphi .$$

From this relation, we compute the distance a of the shear center from the x-axis

$$a = \frac{2r}{\pi} \left. (\sin \varphi) \right|_{-\frac{\pi}{2}}^{\frac{\pi}{2}} = \frac{4r}{\pi} = 1,273 \, r .$$

Alternatively, we may compute this value directly from Eq. (3.41)$_1$

$$y_D = a = \frac{1}{J_{yy}} \int_L a(\zeta) S_y(\zeta) \, d\zeta = \frac{2}{\pi \delta r^3} \int_{-\frac{\pi}{2}}^{\frac{\pi}{2}} r \delta r^2 \cos \varphi \, r \, d\varphi = \frac{4r}{\pi} .$$

3.4 Influence of Distributed Loads

In this section, we will discuss the influence of distributed loads acting on the outer surface of a beam on the shear stresses of this beam.

For the most simple case of plane bending of a rectangular cross section, the load may be given by functions p_x, p_z acting on the upper surface of an element of length dx. The equivalent distributed forces (and moments) $n(x)$, $q(x)$ and $m(x)$, respectively, acting on the beam axis are

$$n(x) = p_x b, \quad m(x) = -p_x b \frac{h}{2}, \quad q(x) = p_z b. \tag{3.42}$$

From the differential equation for the resultants $N(x)$, $Q(x)$ and $M(x)$, i.e. from

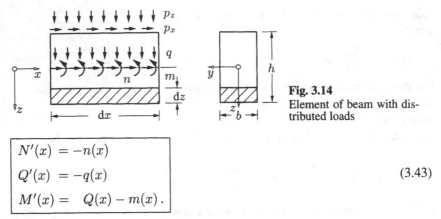

Fig. 3.14
Element of beam with distributed loads

$$
\begin{aligned}
N'(x) &= -n(x) \\
Q'(x) &= -q(x) \\
M'(x) &= Q(x) - m(x).
\end{aligned}
\tag{3.43}
$$

these resultants can easily be determined. However, the question remains, whether such distributed forces will influence the shear stresses.

1. Influence of a distributed axial load p_x differential equation

Since a symmetric rectangular cross section is considered, we can make the same assumptions as before, namely Eqs. (3.3) and (3.25). As before a small element of length dx and height dz may be cut out of the beam. Again integrating over the relevant surfaces, equilibrium of the forces in the x direction yields (Eq. 3.26)

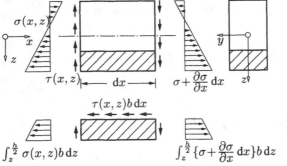

Fig. 3.15
Small element of a beam with distribution of stresses

$$
\tau(z)\, b \, dx = \int_z^{\frac{h}{2}} \frac{\partial \sigma}{\partial x}\, dx\, b\, dz .
\tag{3.44}
$$

The partial differential of σ is (Eq. 3.13)

$$
\frac{\partial \sigma}{\partial x} = \frac{\partial}{\partial x}\left\{ \frac{N(x)}{A} + \frac{M(x)}{J} z \right\} = \frac{1}{A}\frac{dN}{dx} + \frac{dM}{dx}\frac{z}{J}
\tag{3.45}
$$

from which, with Eq. (3.43)

$$
\frac{\partial \sigma}{\partial x} = -\frac{n(x)}{A} + \frac{z}{J}\left[Q(x) - m(x) \right].
\tag{3.46}
$$

Introducing this relation into (3.44), we find

$$\tau(z) = [Q(x) - m(x)] \frac{S_y(z)}{Jb} - \frac{1}{b} \frac{n(x)}{A} \int\limits_{z}^{\frac{h}{2}} b\,dz.$$

(3.47)

The integral herein represents the zeroth moment of the shaded part of the cross-sectional area with respect to the y axis.

For the rectangular cross section under consideration,

$$\frac{n}{A} \int\limits_{z}^{\frac{h}{2}} b\,dz = \frac{n}{2}\left(1 - \frac{2z}{h}\right) = p_x \frac{b}{2}\left(1 - \frac{2z}{h}\right).$$

(3.48)

Thus, with (3.31), we get

$$\tau(z) = \frac{3}{2}\frac{Q}{A}\left\{1 - \left(\frac{2z}{h}\right)^2\right\} + \frac{p_x}{4}\left\{1 + \frac{4z}{h} - 3\left(\frac{2z}{h}\right)^2\right\}.$$

(3.49)

The result shows that the additional term of the shear stress also varies parabolically with z. The stress is zero when $z = h/2$, and $-p_x$ when $z = -h/2$, which coincides with the load on the upper surface, because of $\sigma_{xz} = \sigma_{zx}$.

This result also remains valid when we leave rectangular cross sections, and deal with more general, but still symmetric cross sections, having a symmetry with respect to the loading plane, i.e. the xz plane.

2. Influence of a distributed vertical load p_z

In the preceding considerations, we have discussed the influence of distributed horizontal loads on the shear stresses. In the following paragraph, we will examine the influence of vertical loads p_z on the classical stress distributions. If again symmetric rectangular cross sections are considered first, we can take most of the assumptions as before, i.e. we introduce Eqs. (3.13), (3.25) and (3.30) to describe the stress components σ_{xx}, σ_{xy} and σ_{xz}. Since, however, now $\sigma_{zz} = -p_z$ when $z = -h/2$, assumption (v) (Eq. 3.3) has to be modified in such a way that only

$$\sigma_{yy} = \sigma_{yz} = 0.$$

(3.50)

We note, that via Hooke's law any assumption $\sigma_{zz} \neq 0$ would influence the magnitude of the stress σ_{xx} in axial direction. This influence, however, will be neglected here. Insofar, this assumption, too, is inconsistent with Hooke's law.

As before a small element of length dx and height dz is cut out of the beam. What differs from the preceding considerations is that we now allow normal stresses σ_{zz} acting in the plane parallel to the neutral surface. Integrating over the relevant surfaces, equilibrium of the forces in z direction yields

$$-\int \tau(x, z)\,b\,dz + \int \left\{\tau(x, z) + \frac{\partial\tau}{\partial x}\,dx\right\}b\,dz - \sigma_{zz}(z)\,b\,dx = 0,$$

from which

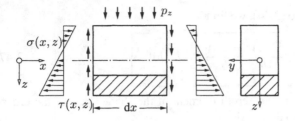

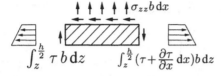

Fig. 3.16
Small element of a beam
with distribution of stresses

$$\sigma_{zz}(z) = \frac{1}{b}\int_z^{\frac{h}{2}} \frac{\partial \tau}{\partial x}\, b\, dz = \frac{1}{b}\int_z^{\frac{h}{2}} \frac{\partial \tau}{\partial x}\, dA\,. \tag{3.51}$$

The partial differential of τ is obtained from Eq. (3.30)

$$\frac{\partial \tau}{\partial x} = \frac{\partial}{\partial x}\left\{\frac{Q(x)S_y(z)}{Jb}\right\} = \frac{dQ}{dx}\frac{S_y}{Jb} = -q\frac{S_y}{Jb}\,. \tag{3.52}$$

Thus, we find

$$\sigma_{zz} = -\frac{q}{b}\int_z^{\frac{h}{2}} \frac{S_y}{J}\, dz\,. \tag{3.53}$$

For a rectangular cross section the integral is

$$\int_z^{\frac{h}{2}} S_y\, dz = \frac{J}{2}\left\{1 - \frac{3}{2}\frac{2z}{h} + \frac{1}{2}\left(\frac{2z}{h}\right)^3\right\}, \tag{3.54}$$

from which we finally get

$$\sigma_{zz} = -\frac{p_z}{2}\left\{1 - \frac{3}{2}\frac{2z}{h} + \frac{1}{2}\left(\frac{2z}{h}\right)^3\right\}. \tag{3.55}$$

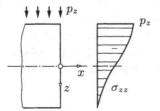

Fig. 3.17
Stresses σ_{zz}

We realize that any distributed vertical load p_z is associated with a normal stress σ_{zz}, and that this stress is of the order of this pressure.

Again, this result also remains valid for any symmetric, but more general cross section.

Example 3.8:
A simply supported beam is subject to a horizontal load on top of the upper surface as shown in the figure. Compute the shear stresses of this beam.

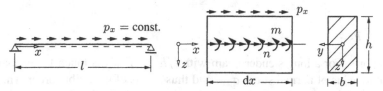

Solution:

The equivalent load with respect to the beam axis from Eq. (3.42) is

$$n = p_x b, \quad m = -p_x b \frac{h}{2}.$$

With this information, we can solve the differential equations for the resultants (3.43), which - taking into account the boundary conditions

$$M(0) = M(l) = 0, \quad N(l) = 0$$

have the following solutions

$$N(x) = p_x(l - x) b,$$

$$Q(x) = Q_0 = -p_x \frac{h}{2} b,$$

$$M(x) \equiv 0.$$

Now introducing this information into (3.47), we arrive at

$$\tau(z) = -\frac{p_x}{2} \left\{ 1 - \frac{2z}{h} \right\},$$

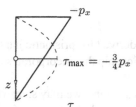

Fig. 3.18 Shear stresses

a linear function of z. The shear stress is zero when $z = h/2$, and $-p_x$ when $z = -h/2$. It is obvious that this solution deviates considerably from the classical parabolic distribution of the shear stresses, which also has been sketched using a dashed line (with the maximum value of $\tau_{\max} = -3/4 \, p_x$ at $z = 0$).

If now a vertical load p_z is applied to the above beam, we find

$$q = p_z b \quad \rightarrow \quad M(x) = \frac{1}{2} p_z b\, x(l - x),$$

$$Q(x) = \frac{1}{2} p_z b (l - 2x).$$

Thus the (classical) shear stress τ is determined from Eq. (3.17) with a maximum value of

$$\tau_{max} = \frac{3}{4} p_z \frac{l}{h}$$

at both supports. For a long slender beam with $l/h \gg 1$ this is much larger than the maximum value of normal stress σ_{zz}, and thus σ_{zz} is of negligible order. This, however, is not the case in the middle of the beam or, generally, for short beams.

3.5 Stresses in Non-prismatic Beams

The beams discussed in the preceding sections were assumed to be prismatic, i.e. of constant cross section through their length. However, the stress formulas derived for prismatic beams may also be used with good accuracy for a non-prismatic beam, provided that the variation in cross-sectional dimensions are very gradual. For example, the normal stresses due to bending in a bar with a slightly tapered form can be calculated from Eq. (3.13); the result will be accurate within a few percent if the angle of the taper is small, as can be shown by comparison with some results from elasticity.

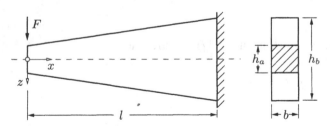

Fig. 3.19 Non-prismatic beam

For the shear stresses, however, the formula (3.32) derived for prismatic bars is no longer adequate. Instead, we must derive a new relationship that incorporates the effect of the changing height (and width) of the beam.

Considering first plane bending of rectangular cross sections, we may start with the same assumptions that have been made for calculating the shear stresses of constant rectangular cross sections, and of symmetric general cross sections, respectively, Eqs. (3.3), (3.13), and (3.25). We note that for changing widths assumption (3.25) has to be replaced by (3.34) again.

In the same way as we did for prismatic beams, we cut a small element of length dx out of the non-prismatic beam (Fig. 3.20) with

$$2 \tan \alpha = \frac{\partial h(x)}{\partial x} = h'(x)$$

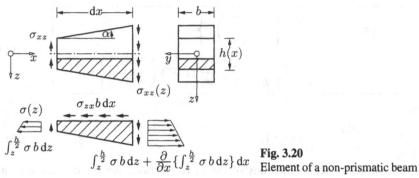

Fig. 3.20
Element of a non-prismatic beam

Integrating over the relevant surfaces, the equilibrium of the forces in x direction yields

$$\sigma_{xz}(z) = \frac{1}{b} \frac{\partial}{\partial x} \left\{ \int_z^{h(x)/2} \sigma b \, dz \right\}. \tag{3.56}$$

Introducing here Eq. (3.13), we arrive at

$$\sigma_{xz}(z) = \frac{1}{b} \frac{\partial}{\partial x} \left\{ \frac{N}{A(x)} b \int_z^{h(x)/2} z \, dz + \frac{M(x)}{J(x)} b \int_z^{h(x)/2} z \, dz \right\} \tag{3.57}$$

and finally

$$\boxed{\sigma_{xz}(z) = \frac{Q S_y}{J b} + \frac{1}{b} \left\{ N \left(\frac{A^*}{A} \right)' + M \left(\frac{S_y}{J} \right)' \right\},} \tag{3.58}$$

where

$$A^*(x, z) = \int_z^{h(x)/2} b \, dz, \qquad S_y(x, z) = \int_z^{h(x)/2} z b \, dz \tag{3.59}$$

are the zeroth and the first moment, respectively, of the shaded portion of the cross-sectional area with respect to the y axis, which are now functions of x and z.

For a rectangular cross section, we find

$$A^*(x, z) = \frac{1}{2} A(x) \left\{ 1 - \frac{2z}{h(x)} \right\}$$

$$S_y(x, z) = \frac{1}{8} h(x) A(x) \left\{ 1 - \left(\frac{2z}{h(x)} \right)^2 \right\} \tag{3.60}$$

$$\left(\frac{A^*}{A} \right)' = \frac{2z}{h^2} \tan \alpha$$

$$\left(\frac{S_y}{J} \right)' = -\frac{3}{h^2} \left\{ 1 - 3 \left(\frac{2z}{h(x)} \right)^2 \right\} \tan \alpha$$

and thus

$$\boxed{\begin{aligned} \sigma_{xz}(x, z) = {}& \frac{3}{2} \frac{Q(x)}{A(x)} \left\{ 1 - \left(\frac{2z}{h} \right)^2 \right\} \\ &+ \left\{ \frac{N(x)}{A(x)} \frac{2z}{h} - \frac{M(x)}{2W(x)} \left[1 - 3 \left(\frac{2z}{h} \right)^2 \right] \right\} \tan \alpha , \end{aligned}}$$

(3.61)

with

$$W = 2 \frac{J(x)}{h(x)} = \frac{1}{6} bh^2(x) ,$$

(3.62)

the well-known section modulus of the cross section. We realize that the shear stresses on the cross section are dependent not only upon the shear force Q, but also upon the normal force N and the bending moment M, and the rate of change of h with respect to x.

Example 3.9:
As a specific example, we investigate the distribution of shear stresses for the cantilever beam of rectangular cross section shown in Fig. 3.19. The beam has heights h_a and $h_b = 2h_a$ at its ends, and a uniform taper

$$h(x) = h_a \left(1 + \frac{x}{l} \right), \quad h'(x) = 2 \tan \alpha = \frac{h_a}{l} .$$

Solution:

We find

$$N \equiv 0, \quad Q = -F, \quad M = -Fx .$$

Thus, at both ends, and at the middle of the beam ($x = l/2$), the following distributions are obtained

$$x = 0 : \quad \sigma_{xz} = -\frac{3}{2} \frac{F}{A_a} \left\{ 1 - \left(\frac{2z}{h_a} \right)^2 \right\},$$

$$x = l/2 : \quad \sigma_{xz} = -\frac{2}{3} \frac{F}{A_a} = -\frac{F}{A_m},$$

$$x = l : \quad \sigma_{xz} = -\frac{3}{8} \frac{F}{A_a} \left\{ 1 + \left(\frac{2z}{h_b} \right)^2 \right\},$$

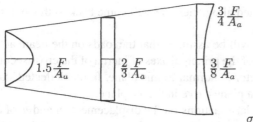

Fig. 3.21
Stresses in a non-prismatic
σ_{xz} beam

The different distributions are plotted in Fig. 3.21. At the left end, the moment M
is zero, and hence we find the same distribution of the shear stress as in a prismatic
beam, with the maximum value

$$|\sigma_{xz\max}| = \frac{3}{2} \frac{F}{A_a}$$

at the neutral axis ($z = 0$).

At the middle of the beam, we obtain the interesting result that the shear stress
is uniformly distributed over the height of the beam. At the right end of the beam,
again a parabolic distribution of the shear stress is obtained. Note, however, that the
maximum value

$$|\sigma_{xz\max}| = \frac{3}{4} \frac{F}{A_a}$$

occurs at the outer edges of the beam.

□

If in addition to the preceding analysis a changing width is examined, a similar
calculation leads us to additional information about shear stress σ_{xy}, namely

$$\boxed{\sigma_{xy}(x, z) = \sigma_{xy}(x, z) \frac{y}{b(x)} \tan \beta \,,} \tag{3.63}$$

while the relation for σ_{xz} remains unaltered, and

$$\tan \beta = \frac{1}{2} b'(x) \tag{3.64}$$

expresses the rate of change of the width with respect to x.

3.6 Deflections of Beams

The deflections of a loaded beam are analyzed by considering the deflections which
result from the action of axial forces, bending moments, and shear forces. The total
deflection is then obtained by adding the different contributions. The more slender the beam, the more predominant is the part of deflection due to the bending
moments, and the more negligible is the part due to the shear forces. In many practical applications, therefore, in accord with Bernoulli's hypothesis the effect of these

forces on the deflection is neglected altogether. We will come back to this point in Chapter 6.

In the following discussion it will be assumed that the loads on the beam act in the xz plane, which contains one of the principal axes of inertia of the cross section, and that the bending moment vector is normal to this plane. In the deflected state, the axis of the beam will then be a plane curve in the xz plane.

In order to derive the differential equations for the displacements u and w of the beam axis, we will utilize the relationships between these functions and the radius of curvature R, and the strain in axial direction ε_0, respectively, i.e.

$$\varepsilon_0 = u'(x) \tag{3.65}$$

and

$$\frac{1}{R} = \frac{-w''(x)}{(1 + w'^2)^{3/2}}, \tag{3.66}$$

(see Eqs. 3.4 and 3.12).

In many practical cases, the deflections of a beam are small compared to its span, and thus the term $w'^2(x)$ may be neglected against unity. Then Eq. (3.66) assumes the simplified form of Eq. (3.4)

$$\frac{1}{R} = -w''(x). \tag{3.67}$$

In Section 3.1 relations between ε_0 and $1/R$ and the stress resultants N and M, respectively, have been determined for isothermal processes (see Eq. 3.12). In the sequel, moreover, the influence of a temperature field varying linearly with z on these relations is considered. We thus have (see Eq. 2.32$_1$)

$$\sigma_{xx} = E\varepsilon_{xx} - E\alpha\Theta, \tag{3.68}$$

where the increase in temperature is given as

$$\Theta = \theta_0 + \theta_1 z. \tag{3.69}$$

Defining now

$$N^\theta = E\alpha \int_A \Theta \, dA, \quad M^\theta = E\alpha \int_A z\Theta \, dA, \tag{3.70}$$

we arrive at

$$N = EA\varepsilon_0 - N^\theta, \quad M = \frac{EJ}{R} - M^\theta, \tag{3.71}$$

wherein

$$N^\theta = EA\alpha\theta_0, \quad M^\theta = EJ\alpha\theta_1 \tag{3.72}$$

describe the contributions of the temperature field. Introducing these informations into Eqs. (3.65) and (3.67), we finally get the differential equations

$$\boxed{\begin{aligned} EAu'(x) &= N(x) + N^\theta \\ EJw''(x) &= -M(x) - M^\theta. \end{aligned}} \tag{3.73}$$

Differentiating these equations with respect to x and then substituting Eqs. (3.43), yields

$$
\begin{aligned}
[EAu'(x)]' &= -n(x) \\
[EJw''(x)]'' &= q(x) + m'(x)
\end{aligned}
\tag{3.74}
$$

The solutions of these differential equations - for constant values of EA and EJ - consist of successive integrations, with the resulting constants of integration being evaluated from the boundary conditions of the beam. Some typical examples of boundary conditions are given in Table 3.1.

system	kinematical conditions	dynamical conditions
	$u = 0; \quad w = 0$	$- \qquad M = 0$
	$- \qquad w = 0$	$N = 0; \quad M = 0$
	$u = 0; \quad w' = 0$	$- \qquad Q = 0$
	$u = 0; \quad w = 0, \; w' = 0$	$- \qquad -$
	$- \qquad w = 0, \; w' = 0$	$N = 0 \qquad -$
	$- \qquad -$	$N = 0; \quad Q = 0, \; M = 0$

Table 3.1 Boundary conditions of beams

From the derivation of the equations, we see that they are valid only when Hooke's law applies for the material and when the slopes of the deflection curve are very small. Also, it should be realized that Eqs. (3.73) need some information about the resultants $N(x)$ and $M(x)$, respectively, whereas Eqs. (3.74) do not. From this we can conclude that the applicability of the former equations is more suitable to statically determinate systems. The latter set of equations may also be used to determine the deflections - and thus the redundants - of statically indeterminate systems.

Example 3.10:

The frame has a constant flexural rigidity EJ. Compute the deflection curves $w_1(x_1)$ and $w_2(x_2)$, the reactions as well as the normal force, the shear force, and the bending moment diagrams.

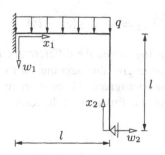

Solution:

We divide the frame into two regimes and introduce for each a coordinate system as shown in the diagram.

The system is statically indeterminate of degree one. Thus, since the function of the bending moment is a priori not known, we start from Eq. (3.74), neglecting here for the sake of simplicity the influence of the normal forces on the deflections. We therefore have the differential equations

$$EJw_1''''(x_1) = q, \quad \text{(in regime 1)}$$
$$EJw_2''''(x_2) = 0, \quad \text{(in regime 2)}$$

and pertinent boundary conditions

$$w_1(0) = w_1'(0) = 0, \quad Q_1(l) = -EJw_1'''(l) = 0,$$
$$w_2(0) = w_2(l) = 0, \quad M_2(0) = -EJw_1''(0) = 0,$$

and transition conditions

$$w_1'(l) = w_2'(l), \quad M_1(l) = -M_2(l).$$

These are all together eight conditions to determine the $2 \times 4 = 8$ integration constants of the problem.

The solutions of the above differential equations are

$$EJw_1 = \frac{1}{24} qx_1^4 + \frac{1}{6} c_1 x_1^3 + \frac{1}{2} c_2 x_1^2 + c_3 x_1 + c_4,$$

$$EJw_2 = \frac{1}{6} c_5 x_2^3 + \frac{1}{2} c_6 x_2^2 + c_7 x_2 + c_8.$$

From the boundary and transition conditions, we compute

$$c_3 = c_4 = c_6 = c_8 = 0,$$

and

$$c_1 = -ql, \quad c_2 = \frac{3}{8} ql^2,$$
$$c_5 = \frac{1}{8} ql, \quad c_7 = -\frac{1}{48} ql^3.$$

Thus we arrive at the deflection curves

$$EJw_1 = \frac{q}{48} x_1^2 \left(2x_1^2 - 8lx_1 + 9l^2\right),$$

$$EJw_2 = \frac{q}{48} lx_2 \left(x_2^2 - l^2\right).$$

Having solved the differential equations for the deflections, we are now in the position to give the diagrams for the stress resultants, although the system is statically indeterminate. We therefore refer to Eqs. (3.43) and (3.73) from which we can compute the functions of the resultants as derivatives of these deflections.

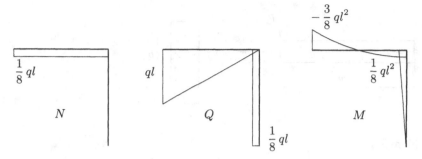

We mention in passing that without recourse to any theory of statically indeterminate systems, we nevertheless were able to solve a statically indeterminate
system. Thus Eqs. (3.74) moreover deliver a first method for solving statically indeterminate systems.

3.7 Exercises to Chapter 3

Problem 3.1:

For the following cross sections, compute the principal moments of inertia and
planes:

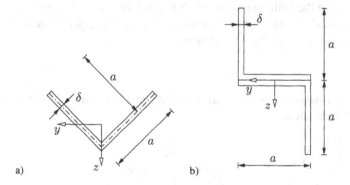

Problem 3.2:

For the cross section shown in the figure
and given dimensions $h = 30$ cm, $b =
10$ cm, $d = 1$ cm, $t = 1.6$ cm, determine
the distance a such that the principal moments of inertia J_{yy} and J_{zz} are equal.

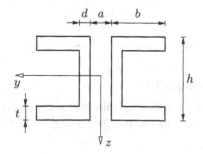

Problem 3.3:

For the cantilever beam shown below and for $a = b/4$, determine using the maximum normal stress criterion the equivalent stress at $x = 0$.

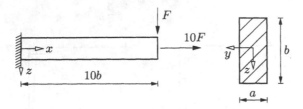

Problem 3.4:

Determine the maximum compressive and tensile stresses in the Z section of the beam shown in the figure, locate the neutral axis, and draw a diagram to show the stress distribution. All lengths are in cm.

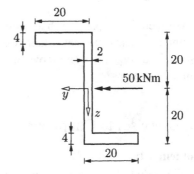

Problem 3.5:

A cantilever beam with the built-up section of Problem 3.1b is at its free end ($x = l$) subjected to a normal (tensile) force F acting on top of the upper vertical leg. Determine the extreme values of the normal stresses.

Problem 3.6:

For the following thin-walled cross sections, compute the shear stress distribution, and the maximum shear stress.

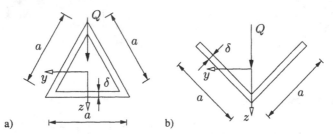

Problem 3.7:

For the following thin-walled cross sections, compute the shear stress distribution, the maximum shear stress, and locate the shear center.

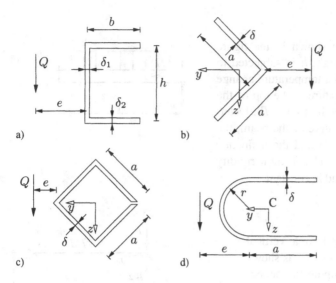

a)   b)

c)   d)

Problem 3.8:

For the frames below with constant flexural rigidity EJ, compute the deflection curves $w_1(x_1)$ and $w_2(x_2)$.

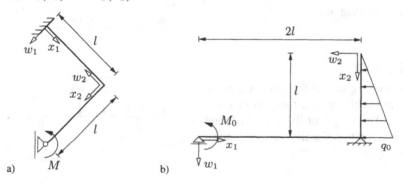

a)   M   b)

Problem 3.9:

For the given frame, compute the deflection curves $w_1(x_1)$ and $w_2(x_2)$ and the reactions at the shafts after an increase in temperature of θ_0. The length of each beam is l, the flexural rigidity is EJ, and the axial rigidity is EA.

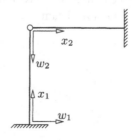

Problem 3.10:

The cantilever beam shown in the figure with length l and height h is subjected to a vertical load q and a temperature change such that the temperature at the top of the beam is T_1 and at the bottom is T_2. Determine the deflection curve of the beam, the angle of rotation $w'(l)$ and the deflection $w(l)$ at the free end. The flexural rigidity of the beam is EJ and is constant.

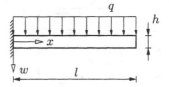

Problem 3.11:

A frame is subjected to a vertical force F at the point $x_1 = l$ as shown in the diagram. Compute the deflection curves $w_1(x_1)$ and $w_2(x_2)$. Note that the flexural rigidity of the horizontal section of the frame is two times that of the angled-section as shown in the diagram.

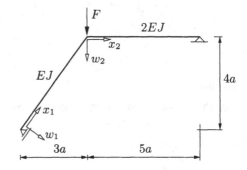

Problem 3.12:

The flexural rigidity of the cantilever beam shown in the figure is a function of coordinate x. The length of the beam is l, and it is subjected to a linearly varying load of intensity $q(x) = \alpha x + \beta$. For

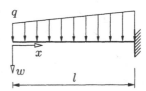

$$EJ(x) = k\left(\frac{\alpha}{6}x^3 + \frac{\beta}{2}x^2\right),$$

determine the deflection curve of the beam, the angle of rotation, and the deflection at the free end of the beam.

4. Torsion of Prismatic Bars

4.1 Solid Cross Sections

Torsion of a uniform bar of homogeneous isotropic material means twisting deformation produced by equal and opposite twisting moments applied to the ends of the
bar.

The problem of uniform torsion for bars of any cross section is solved by the
St. Venant semi-inverse method. Certain features of the solution are at first assumed,
but the fact that it is then possible to satisfy all the equations and boundary conditions validates the assumptions.

It is assumed first that the cross sections rotate about the x axis, the (small)
angle of rotation being ϑx, without distortion in their own planes, and where ϑ is
the angle of twist per unit length, or short, the twist.[1] The displacement components
u_y, u_z in the plane then correspond to a small rotation ϑx of a rigid lamina, and are

$$u_y = -\vartheta xz, \quad u_z = \vartheta xy. \tag{4.1}$$

The axial, or warping, displacement u_x is taken as a function of y and z.

$$u_x = \vartheta \, \psi(y, z), \tag{4.2}$$

where $\psi(y, z)$ is the warping function.

The six strain components follow from these displacements as

$$\varepsilon_{xx} = \frac{\partial u_x}{\partial x} = 0, \quad \varepsilon_{yy} = \frac{\partial u_y}{\partial y} = 0, \quad \varepsilon_{zz} = \frac{\partial u_z}{\partial z} = 0 \tag{4.3}$$

and

$$\varepsilon_{xy} = \frac{\vartheta}{2} \left\{ \frac{\partial \psi}{\partial y} - z \right\},$$

$$\varepsilon_{xz} = \frac{\vartheta}{2} \left\{ \frac{\partial \psi}{\partial z} + y \right\}, \tag{4.4}$$

$$\varepsilon_{yz} = \frac{1}{2} \left\{ \frac{\partial u_z}{\partial y} + \frac{\partial u_y}{\partial z} \right\} = 0.$$

---

[1] We note that according to the assumptions of St. Venant's theory this twist has to be a constant. For non-uniform torsion, where the rotation of an element of length dx is described
by $d\varphi = \vartheta(x) \, dx$, we refer to Section 4.4.

These give, with Hooke's law, the stress components

$$\sigma_{xx} = \sigma_{yy} = \sigma_{zz} = \sigma_{yz} = 0 \tag{4.5}$$

$$\sigma_{xy} = G\vartheta \left\{ \frac{\partial \psi}{\partial y} - z \right\},$$

$$\sigma_{xz} = G\vartheta \left\{ \frac{\partial \psi}{\partial z} + y \right\}. \tag{4.6}$$

From the three equations of equilibrium (Eqs. 1.61 and 1.62), two are identically satisfied, and the third reduces to

$$\frac{\partial \sigma_{yx}}{\partial y} + \frac{\partial \sigma_{zx}}{\partial z} = 0, \tag{4.7}$$

provided, we neglect acceleration and body forces.

The problem of torsion may now be treated in different ways:

1. The first system is obtained, when we introduce Eqs. (4.6) into the equilibrium equation (4.7)

$$\boxed{\Delta \psi = \frac{\partial^2 \psi}{\partial y^2} + \frac{\partial^2 \psi}{\partial z^2} = 0,} \tag{4.8}$$

which gives an equation for the warping function.

The boundary conditions of the problem are such that the resultant shear stress at the boundary must be directed along the boundary curve, since the surface is assumed to be stress free, i.e.

$$\sigma_{xy} \frac{dz}{ds} - \sigma_{xz} \frac{dy}{ds} = 0, \tag{4.9}$$

where the boundary itself is introduced as

$$y = y(s), \quad z = z(s) \tag{4.10}$$

with arc length s measured from some fixed point. Using Eqs. (4.6), the boundary condition may be expressed in terms of the warping function as

$$\frac{\partial \psi}{\partial n} = \frac{1}{2} \frac{d}{ds} \left[y^2(s) + z^2(s) \right] = \frac{1}{2} \frac{d}{ds} r^2(s) \tag{4.11}$$

2. For the second way, we eliminate $\psi(y, z)$ from the two equations in (4.6) to obtain

$$\frac{\partial \sigma_{xy}}{\partial z} - \frac{\partial \sigma_{xz}}{\partial y} = -2G\vartheta. \tag{4.12}$$

It follows now that σ_{xy} and σ_{xz} may be expressed in terms of a single function $T(y, z)$, Prandtl's stress function, as

$$\sigma_{xy} = 2G\vartheta \frac{\partial T}{\partial z}, \quad \sigma_{xz} = -2G\vartheta \frac{\partial T}{\partial y}. \tag{4.13}$$

This function must satisfy

$$\boxed{\Delta T = \frac{\partial^2 T}{\partial y^2} + \frac{\partial^2 T}{\partial z^2} = -1} \qquad (4.14)$$

with boundary condition (4.9) now taking the form

$$\frac{\partial T}{\partial s} = \frac{\partial T}{\partial y}\frac{dy}{ds} + \frac{\partial T}{\partial z}\frac{dz}{ds} = 0. \qquad (4.15)$$

This means that T must be constant around the entire closed boundary curve, with different constants at different curves, e.g. if there are holes. One of these constants may be arbitrarily assigned as zero, such that

$$\Gamma : \ T(s) = 0 \qquad (4.16)$$

for bodies without holes.

It is advantageous to consider, in general terms, the evaluation of the torque M_T and the warping displacement $\vartheta\psi$:

The torque M_T is given by

$$M_T = \int_A (\sigma_{xz} y - \sigma_{xy} z)\, dA \qquad (4.17)$$

the integral over the material cross section (excluding holes). This can be expressed in terms of the stress function T as

$$M_T = -2G\vartheta \int_A \left\{ \frac{\partial T}{\partial y} y + \frac{\partial T}{\partial z} z \right\} dy\, dz\,. \qquad (4.18)$$

Finally, after integrating by parts, the total torque is given by

$$M_T = 4G\vartheta \int_A T(y, z)\, dA\,, \qquad (4.19)$$

which is twice the volume under the $T(y, z)$ surface multiplied by the factor $2G\vartheta$. From this relation, we can introduce the polar moment of inertia J_T as

$$J_T = 4 \int_A T(y, z)\, dA\,. \qquad (4.20)$$

The increment of $\psi(y, z)$ corresponding to any arc element ds in the material section is

$$d\psi = \frac{\partial\psi}{\partial y}\, dy + \frac{\partial\psi}{\partial z}\, dz\,, \qquad (4.21)$$

i.e. from Eq. (4.6)

$$d\psi = \left\{ \frac{\sigma_{xy}}{G\vartheta} + z \right\} dy + \left\{ \frac{\sigma_{xz}}{G\vartheta} - y \right\} dz$$

$$= \frac{1}{G\vartheta}(\sigma_{xy}\, dy + \sigma_{xz}\, dz) + (z\, dy - y\, dz)\,. \qquad (4.22)$$

The sum of these increments for all arc elements of a closed curve lying entirely in the material must vanish to avoid discontinuity in ψ. Thus, we find

$$\int_C \tau_s \, ds = 2G\vartheta A_C , \tag{4.23}$$

the shear-circulation theorem for a closed circle C, where A_C is the area enclosed by C, and τ_s is the shear component in the direction of s.

Example 4.1:

Determine the stresses and the warping ψ of a solid elliptical cross section.

Solution:

The problem is solved (indirectly) by finding the stress function T which satisfies within the ellipse

$$\frac{y^2}{a^2} + \frac{z^2}{b^2} = 1$$

the differential equation (4.14), and on the ellipse satisfies the boundary condition $T = 0$. The function is

$$T = m\left\{ \frac{y^2}{a^2} + \frac{z^2}{b^2} - 1 \right\}, \quad m = -\frac{1}{2}\frac{a^2 b^2}{a^2 + b^2} .$$

The stress components are (Eq. 4.13)

$$\sigma_{xy} = 2G\vartheta m \frac{2z}{b^2} , \quad \sigma_{xz} = -2G\vartheta m \frac{2y}{a^2} .$$

The shear stress τ is the resultant of σ_{xy} and σ_{xz}, and so

$$\tau = 4G\vartheta m \sqrt{\frac{y^2}{a^4} + \frac{z^2}{b^4}}$$

with the maximum value

$$\tau_{max} = 2G\vartheta \frac{a^2 b}{a^2 + b^2}$$

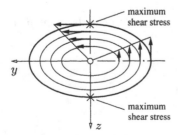

maximum shear stress

maximum shear stress

Fig. 4.1
Stresses in an elliptical cross section

on the boundary at the ends of the minor axis, that is, at the points nearest the axis of the torsional rotations.

From Eq. (4.6), we determine by integration the function $\psi(y, z)$

$$\psi = -\frac{a^2 - b^2}{a^2 + b^2} yz.$$

The lines of constant values of the warping are shown in Fig. 4.2.

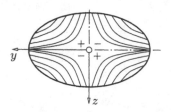

Fig. 4.2
Warping of an elliptical cross section

The torque is, from Eq. (4.19)

$$M_T = G\vartheta \frac{\pi a^3 b^3}{a^2 + b^2}.$$

From these results, we can deduce the following relations

$$
\begin{array}{ll}
\tau_{\max} = \dfrac{M_T}{W_T}, & W_T = \dfrac{\pi}{2} ab^2, \\[3mm]
\vartheta = \dfrac{M_T}{GJ_T}, & J_T = \pi \dfrac{a^3 b^3}{a^2 + b^2}.
\end{array}
$$

□

Corresponding values of the polar moment of inertia J_T and the section modulus W_T may be determined for cross sections of different kinds:

1. Equilateral triangle

Let the boundary of a torsion member be an equilateral triangle with $h = \sqrt{3}a/2$.

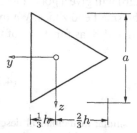

Fig. 4.3
Equilateral triangle

Proceeding as for the elliptical cross section, we find

$$
\begin{array}{ll}
\tau_{\max} = \dfrac{M_T}{W_T}, & W_T = \dfrac{1}{20} a^3, \\[3mm]
\vartheta = \dfrac{M_T}{GJ_T}, & J_T = \dfrac{\sqrt{3}}{80} a^4.
\end{array}
\tag{4.24}
$$

2. Narrow rectangular cross section

Consider a bar subjected to torsion. Let the cross section of the bar be a solid rectangle with width b and depth h, where $b \ll h$. From the different analogies (e.g. the soap-film analogy, originally proposed by L. Prandtl.),[2] we may conclude that except for the region near $z = \pm h/2$ the stress components σ_{xy} and σ_{xz} are ap-

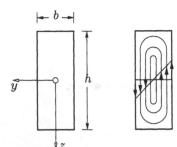

Fig. 4.4
Narrow rectangle

proximately independent of z, and

$$\sigma_{xy} \cong 0, \quad \sigma_{xz} \cong \tau(y) = 2G\vartheta y .$$

(4.25)

Thus, from Eq. (4.13), we determine

$$T \cong T(y) = \frac{b^2}{8}\left\{1 - \left(\frac{2y}{b}\right)^2\right\},$$

(4.26)

and furthermore

$$W_T = \frac{1}{3} hb^2, \quad J_T = \frac{1}{3} hb^3 .$$

(4.27)

We note, however, that near the ends, of course, these results, which are valid only for narrow cross sections, do not apply. The exact theory for the rectangle is then required.

The simple parabolic approximate form of T (Eq. 4.26) will give a good approximation, since it differs from the true solution only in the small end zones. We may generalize Eq. (4.27) by introducing correction factors α and β, as functions of the ratio h/b (see Table 4.1), to give

$$W_T = \alpha \frac{1}{3} hb^2, \quad J_T = \beta \frac{1}{3} hb^3 .$$

(4.28)

---

[2] Analogies exist where physically different problems have similar mathematical descriptions. In this case solutions - or experimental findings - from one problem may be transferred to the other - analogous - problem. The most known analogy to the torsion problem of prismatic bars is that of a membrane (soap film) fixed on a closed boundary, having the same shape as the cross section of the torsion bar, where pressure is applied to one side of the membrane. We therefore refer e.g. to the textbook of Boresi, Schmidt & Sidebottom, *Advanced Mechanics of Materials*, 5th. Edition, John Wiley & Sons, N.Y. etc., 1993.

h/b	1	1.5	2	3	4	8	∞
α	0.63	0.69	0.74	0.80	0.85	0.92	1
β	0.42	0.59	0.69	0.79	0.84	0.92	1

Table 4.1 Correction factors α and β

4.2 Thin-Walled Closed Cross Sections

In the preceding section, we have discussed torsion of prismatic bars with solid cross sections. In the following sections, we now will examine this problem with thin-walled cross sections.

We maintain the assumptions of St. Venant's theory about the displacement components (Eqs. 4.1, 4.2), and furthermore - for the moment - assume unrestrained warping $\psi(y, z)$ in the x direction. In Section 3.2, we discussed beams of thin-walled cross sections subject to shear forces. From this section, we take the modified description of thin-walled cross sections, with cross-sectional centerline (middle line), and additional rectangular coordinates ζ, along the centerline and η, perpendicular to ζ. Thus, the thin-walled cross section is described by the coordinates $y(\zeta), z(\zeta)$ of the centerline, and thickness $\delta(\zeta)$ of the profile.

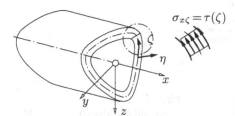

$$\sigma_{x\zeta} = \tau(\zeta)$$

Fig. 4.5
Closed cross section

This allows us to assume

$$\sigma_{x\eta} = 0, \quad \sigma_{x\zeta} = \tau(\zeta) \tag{4.29}$$

for the shear stresses, and

$$\mathbf{u} = \mathbf{u}(x, \zeta) \tag{4.30}$$

for the displacements, constant in the η direction. We note that assumptions (4.29) coincide with the assumptions (3.37) for the shear stresses due to shear forces (Section 3.2).

Defining now the shear flow $t(\zeta)$ as

$$t(\zeta) = \int_{-\delta/2}^{\delta/2} \sigma_{x\zeta}\, \mathrm{d}\eta\,, \tag{4.31}$$

we obtain with $(4.29)_1$

$$t(\zeta) = \tau(\zeta)\delta(\zeta)\,. \tag{4.32}$$

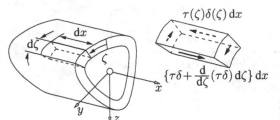

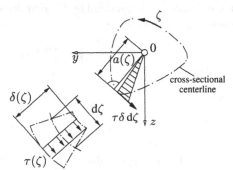

Fig. 4.6
Element of thin-walled tube

From the equilibrium (in axial direction) of a small element cut from the thin-walled tube (Fig. 4.6), we see that

$$t = \tau(\zeta)\delta(\zeta) = \text{const.} , \tag{4.33}$$

provided there are no axial stresses σ_{xx}. Thus, the largest shear stress occurs where the thickness is smallest, and vice versa. Of course, if the thickness is uniform, then the shear stress τ is constant around the tube.

In order to relate the shear flow to the torque M_T acting on the tube, consider an element of length $d\zeta$ in the cross section:

Fig. 4.7
Shear force acting on an element

The total shear force acting on the element is $t\,d\zeta$, and the moment of this force about any point 0 is

$$dM_T = a(\zeta)t\,d\zeta , \tag{4.34}$$

in which $a(\zeta)$ is the distance from 0 to the tangent to the centerline. The total torque then is

$$M_T = t \oint a(\zeta)\,d\zeta = 2A_m t , \tag{4.35}$$

where the integral represents double the area A_m enclosed by the centerline of the tube. From this equation, we find

$$\boxed{t = \tau(\zeta)\delta(\zeta) = \frac{M_T}{2A_m} ,} \tag{4.36}$$

and finally,

$$\tau_{\max} = \frac{M_T}{W_T}, \quad W_T = 2A_m\delta(\zeta)_{\min}, \tag{4.37}$$

Bredt's first formula.

To determine a relation between torque M_T and twist ϑ, we start from assumption (4.30), and

$$u_x(\zeta) = \vartheta\psi(\zeta), \quad u_\zeta = \vartheta x a(\zeta). \tag{4.38}$$

With these displacements, we find

$$\varepsilon_{x\zeta}(\zeta) = \frac{1}{2}\left\{\frac{du_x}{d\zeta} + \vartheta a(\zeta)\right\}, \tag{4.39}$$

and thus from Hooke's law

$$\sigma_{x\zeta}(\zeta) = \tau(\zeta) = 2G\varepsilon_{x\zeta}. \tag{4.40}$$

Integrating this expression over the entire length of the centerline yields

$$\oint \tau(\zeta)\,d\zeta = G\left\{\oint du_x + \vartheta \oint a(\zeta)\,d\zeta\right\}. \tag{4.41}$$

The first integral of the right hand side vanishes (continuity of u_x), and thus with Eq. (4.36), we finally arrive at

$$\vartheta = \frac{M_T}{GJ_T}, \quad J_T = 4A_m^2\left\{\oint \frac{d\zeta}{\delta(\zeta)}\right\}^{-1}, \tag{4.42}$$

Bredt's second formula.

If in Eq. (4.41) the integrals are not taken over the entire length of the centerline, e.g. for the first integral of the right hand side

$$\int_0^\zeta du_x = u_x(\zeta) - u_x(0) = \vartheta\{\psi(\zeta) - \psi(0)\}, \tag{4.43}$$

Bredt's second formula is replaced by a slightly different relation

$$\psi(\zeta) = \psi(0) + 2A_m\frac{\displaystyle\int_0^\zeta \frac{d\zeta}{\delta(\zeta)}}{\displaystyle\oint \frac{d\zeta}{\delta(\zeta)}} - \int_0^\zeta a(\zeta)\,d\zeta \tag{4.44}$$

from which the distribution of the warping along the centerline may be calculated.

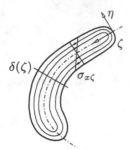

Fig. 4.8
Thin-walled open section

4.3 Thin-Walled Open Sections

Expressions for thin-walled open sections are based on formulas for the torsion of the thin rectangle. For sufficiently thin-walled sections, good approximate formulas are obtained by regarding the stress function T as parabolic across the thickness $\delta(\zeta)$ as in a thin rectangle of the same thickness. From this, it seems reasonable to make the following assumptions as before (cf. Eqs. (4.29)$_1$, (4.30))

$$\text{(i)} \quad \sigma_{x\eta} = 0, \qquad \text{(ii)} \quad \mathbf{u} = \mathbf{u}(x, \zeta).$$

Furthermore, we take $\sigma_{x\zeta}$ from the thin rectangle as a linearly varying function of η. Thus

$$\text{(iii)} \quad t(\zeta) = 0. \tag{4.45}$$

Finally, we assume

$$\text{(iv)} \quad J_T = \gamma \frac{1}{3} \int_L \delta^3(\zeta)\, \mathrm{d}\zeta = \gamma \frac{1}{3} \sum_i \delta_i^3 L_i, \tag{4.46}$$

wherein γ is a form factor (cf. Table 4.2).

Profile	L	⊥	⊏	I
γ	1	1.12	1.12	1.3

Table 4.2 Form factor γ

From these assumptions, similar to the results before, we find

$$\vartheta = \frac{M_T}{G J_T}, \tag{4.47}$$

and

$$\tau_{\max} = \frac{M_T}{W_T}, \qquad W_T = \frac{J_T}{\delta_{\max}}. \tag{4.48}$$

Since according to assumption (iii) the shear flow vanishes, we find for the warping function

$$\psi(\zeta) = \psi(0) - \int_0^\zeta a(\zeta)\,d\zeta.$$ (4.49)

Note that herein the distance $a(\zeta)$ is measured from the elastic line of the bar, i.e. the axis of the shear centers of all cross sections.

The different formulas to describe the torsion of prismatic bars are summarized in Table 4.3.

Example 4.2:

Determine the polar moment of inertia J_T and the section modulus W_T, as well as the warping function $\psi(\zeta)$ of the hollow square cross section of Example 3.6. Compare the different results for closed and open cross sections.

Solution:

The dimensions are taken from Example 3.6 as

$$b = 50, \quad h = 90, \quad \delta = 2 \quad [\text{cm}]$$

and thus we find

$$A_m = bh \qquad\qquad = 4.5 \cdot 10^3 \text{ cm}^2$$
$$W_T = 2bh\delta \qquad\quad = 18 \cdot 10^3 \text{ cm}^3$$
$$\oint \frac{d\zeta}{\delta(\zeta)} = \frac{2}{\delta}(b+h) = 140$$
$$J_T = 2\frac{\delta b^2 h^2}{b+h} \qquad = 578.57 \cdot 10^3 \text{ cm}^4$$

for the closed cross section, and

$$J_T = \frac{2}{3}\delta^3(b+h) = 746.67 \text{ cm}^4$$
$$W_T = \frac{J_T}{\delta} \qquad\quad = 373.33 \text{ cm}^3$$

for the open cross section, where the form factor γ has been introduced as $\gamma = 1$.

Comparing now these results, we get the following ratios

$$J_T \; : \; 3\left\{\frac{bh}{\delta(b+h)}\right\}^2 = 774.87,$$
$$W_T : \; 3\frac{bh}{\delta(b+h)} \qquad = 48.21,$$

which means that the twist ϑ of the open cross section increases by a factor of 774.87, and the maximum shear stress $\tau_{\max}$ by a factor of 48.21, compared with the respective quantities of the closed section.

The distribution of the warping of the closed cross section is calculated from Eq. (4.44)

Twist: $\qquad \vartheta = \dfrac{M_T}{GJ_T} \qquad \rightarrow \qquad$ Rotation: $\quad \varphi = \displaystyle\int_0^L \vartheta(x)\,\mathrm{d}x + \varphi(0)$

Shear stress: $\qquad |\tau|_{max} = \dfrac{|M_T|}{W_T}$

	J_T	W_T	Stresses, form factors
Solid cross sections			
	$\dfrac{\pi d^4}{32}$	$\dfrac{\pi d^3}{16}$	$\sigma_{x\varphi} = \dfrac{M_T}{W_T} r$
	$\dfrac{a^4}{46,19}$	$\dfrac{a^3}{20}$	
	$\beta \dfrac{1}{3} b^3 h$	$\alpha \dfrac{1}{3} b^2 h$	<table><tr><td>h/b</td><td>1</td><td>2</td><td>4</td><td>8</td><td>∞</td></tr><tr><td>α</td><td>0.63</td><td>0.74</td><td>0.85</td><td>0.92</td><td>1</td></tr><tr><td>β</td><td>0.42</td><td>0.69</td><td>0.84</td><td>0.92</td><td>1</td></tr></table>
Thin-walled closed sections			
	$\dfrac{4A_m^2}{\displaystyle\oint \dfrac{\mathrm{d}\zeta}{\delta(\zeta)}}$	$2A_m\delta(\zeta)_{min}$	$\tau(\zeta) = \dfrac{M_T}{2A_m\delta(\zeta)}$
	$\dfrac{\pi}{4} \delta d_m^3$	$\dfrac{\pi}{2} \delta d^2$	
	$\dfrac{2b^2 h^2 \delta}{b + h}$	$2bh\delta$	
Thin-walled open sections			
	$\dfrac{1}{3} \pi d\delta^3$	$\dfrac{J_T}{\delta} = \dfrac{1}{3}\pi d\delta^2$	$|\tau_{max}|(\zeta) = \dfrac{|M_T|}{J_T}\delta(\zeta)_{max}$
	$\gamma \displaystyle\int_L \dfrac{1}{3}\delta^3(\zeta)\,\mathrm{d}\zeta$ $\gamma \displaystyle\sum_i \dfrac{1}{3}\delta_i^3 h_i$	$W_T = \dfrac{J_T}{\delta_{max}}$	<table><tr><td>Profile</td><td>L</td><td>⊥</td><td>[</td><td>I</td></tr><tr><td>γ</td><td>1</td><td>1.12</td><td>1.12</td><td>1.3</td></tr></table>

Table 4.3 Torsion of prismatic bars

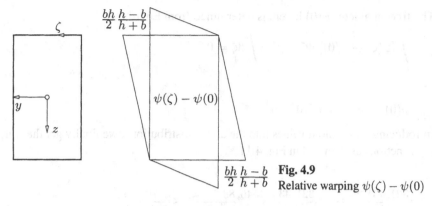

$$\frac{bh}{2}\frac{h-b}{h+b}$$

$\psi(\zeta) - \psi(0)$

$$\frac{bh}{2}\frac{h-b}{h+b}$$ **Fig. 4.9**

Relative warping $\psi(\zeta) - \psi(0)$

$$\psi(\zeta) = \psi(0) + \frac{bh\delta}{b+h}\int_0^\zeta \frac{d\zeta}{\delta(\zeta)} - \int_0^\zeta a(\zeta)\, d\zeta.$$

The free parameter $\psi(0)$ herein is determined from Eq. (4.52)

$$\oint [\psi(\zeta) - \psi(0)]\, d\zeta + \psi(0)\oint d\zeta = 0$$

as

$$\psi(0) = \frac{bh}{4}\frac{h-b}{h+b} = 321.43\ \text{cm}^2.$$

The warping of the open cross section is determined from Eq. (4.49)

$$\psi(\zeta) = \psi(0) - \int_0^\zeta a(\zeta)\, d\zeta,$$

where $a(\zeta)$ is now measured from the shear center of the cross section with

$$y_D = \frac{3b}{2}\frac{h+2b}{h+3b} = 59.38\ \text{cm}.$$

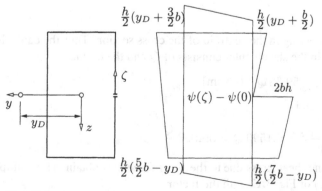

$$\frac{h}{2}\left(y_D + \frac{3}{2}b\right)$$ $$\frac{h}{2}\left(y_D + \frac{b}{2}\right)$$

$\psi(\zeta) - \psi(0)$ $2bh$

$$\frac{h}{2}\left(\frac{5}{2}b - y_D\right)$$ $$\frac{h}{2}\left(\frac{7}{2}b - y_D\right)$$

Fig. 4.10 Relative warping $\psi(\zeta) - \psi(0)$

The free parameter $\psi(0)$ herein is determined from Eq. (4.53)

$$\int_L [\psi(\zeta) - \psi(0)]\, d\zeta + \psi(0) \int_L d\zeta = 0$$

as

$$\psi(0) = bh = 4.5 \cdot 10^3 \ \text{cm}^2.$$

Introducing these initial values into the above distributions, we finally get the warping functions as depicted in Fig. 4.11.

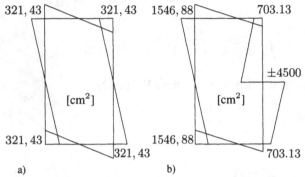

321, 43   321, 43   1546, 88   703.13

± 4500

$[\text{cm}^2]$ $[\text{cm}^2]$

321, 43   1546, 88

321, 43   703.13

a)   b)

Fig. 4.11 Warping function $\psi(\zeta)$ for (a) closed and (b) open cross section

Moreover, from the above result, we can calculate the relative displacement of both ends of the open cross section

$$\Delta u_x = \vartheta \Delta \psi = \frac{M_T}{G J_T} \Delta \psi = 12.05 \frac{M_T}{G} \quad [\text{cm}].$$

Example 4.3:
Determine for the open cross section of Example 3.6 the additional shear stresses due to torsion.

Solution:

The shear force Q is acting in the centroid of the cross section. Thus the equivalent force system acting in the shear center consists of Q and the torque

$$M_T = -Q y_D = -59.38\, Q \quad [\text{kNcm}].$$

From Eq. (4.48), we find

$$\tau_{max}|_{M_T} = \frac{|M_T|}{J_T}\, \delta = 0.1590\, Q = 89.063\, \frac{Q}{A}\,.$$

The distribution of the shear stress due to the shear force Q is obtained by multiplying the function $S(\zeta)$ of Fig. 3.12 with the factor

$$\frac{Q}{J\delta} = \frac{6Q}{\delta^2 h^2(h+3b)},$$

with the maximum value

$$\tau_{max}|_Q = \frac{3}{2}\frac{Q}{\delta h}\frac{h+2b}{h+3b} = 0.0066\,Q = 3.694\,\frac{Q}{A}.$$

It turns out that the maximum shear stress – in the center of the web – consists of two contributions. The first contribution is due to the shear force, and is twice the value of the closed cross section. The second is due to the additional twist about the shear center, and is again by a factor of 24.1 greater than that of the first contribution.

Example 4.4:

Consider a hollow rectangular cross section subject to torsion. The dimensions of this cross section are b, h, and δ_1 and δ_2. Provided there exists a cross section without warping, determine the specific dimensions of this cross section.

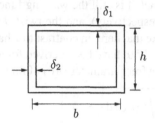

Solution:

The warping of a closed cross section is described by Eq. (4.44) as

$$\psi(\zeta) = \psi(0) + 2A_m \frac{\displaystyle\int_0^\zeta \frac{d\zeta}{\delta(\zeta)}}{\displaystyle\oint \frac{d\zeta}{\delta(\zeta)}} - \int_0^\zeta a(\zeta)\,d\zeta.$$

For the above rectangular cross section, this means

$$\psi(\zeta) - \psi(0) = \frac{bh\delta_1\delta_2}{\delta_1 b + \delta_2 h}\int_0^\zeta \frac{d\zeta}{\delta(\zeta)} - \int_0^\zeta a(\zeta)\,d\zeta = 0.$$

Since this expression has to vanish for every ζ, i.e. every point of the centerline, it follows that

$$\frac{bh\delta_1\delta_2}{\delta_1 b + \delta_2 h}\frac{\zeta}{\delta_2} = \frac{h}{2}\zeta \quad\rightarrow\quad b\delta_1 = h\delta_2.$$

Every closed cross section obeying the relation $h/b = \delta_1/\delta_2$ is free of warping.

4.4 Influence of Restrained Warping

In the uniform torsion of a uniform bar (St. Venant's theory), the originally plane cross sections become warped, the warping displacement u_x being nonzero (unless the section is circular) and such that the cross section becomes a (slightly) curved surface. This warping is the same for all cross sections and is proportional to the twist ϑ, or, equivalently, to the torque M_T. The St. Venant solution prevails only when the warping is free to occur. It needs emendation if the warping is completely or partially prevented, e.g. by an end fitting, or by variation of torque along the length of the bar. As this effect is of great practical importance for structures with thin-walled cross sections, this case will be considered here.

To describe the influence of these constraints on the stress state of the bar, we keep the assumptions of St. Venant's theory about the displacement components (Eqs. 4.1, 4.2 and 4.30), implying

$$\frac{\partial u_y}{\partial x} = -\vartheta z, \quad \frac{\partial u_z}{\partial x} = \vartheta y, \quad \frac{\partial u_x}{\partial x} = \vartheta' \psi(\zeta), \tag{4.50}$$

where $\psi(\zeta)$ is still the warping function of the unrestrained torsion (Eqs. 4.44, and 4.49, respectively), and the twist ϑ is now a function of x, with $(\bullet)' = \mathrm{d}(\bullet)/\mathrm{d}x$. We note that here, in contrast to what has been adopted in Section 4.1, the gradients in Eq. (4.50) have been derived for a small element of length $\mathrm{d}x$.

The free parameter $\psi(0)$ of function $\psi(\zeta)$ may be determined from the condition

$$\boxed{\int_A \psi(\zeta) \, \mathrm{d}A = 0\,,} \tag{4.51}$$

i.e.

$$\oint \psi(\zeta)\delta(\zeta) \, \mathrm{d}\zeta = 0 \tag{4.52}$$

for closed sections, and

$$\int_L \psi(\zeta)\delta(\zeta) \, \mathrm{d}\zeta = 0 \tag{4.53}$$

for open sections. Furthermore, we keep assumption $(4.29)_1$, i.e. $\sigma_{x\eta} = 0$. From Eq. (4.50), it immediately follows that

$$\sigma_{yy} = \sigma_{zz} = \sigma_{yz} = \sigma_{\eta\zeta} = 0 \tag{4.54}$$

and

$$\sigma_{xx} = \sigma(x, \zeta) = E\varepsilon_{xx} = E\vartheta' \psi(\zeta) \tag{4.55}$$

for the axial stresses due to restrained warping. This axial stress must, since the loading is purely torsional, form a zero-force resultant on the cross section

$$N = \int_A \sigma(x, \zeta) \, \mathrm{d}A = E\vartheta' \int_A \psi(\zeta) \, \mathrm{d}A = 0\,, \tag{4.56}$$

which is fulfilled if Eq. (4.51) has been used to determine $\psi(0)$.

The equilibrium equation in axial direction for a small element cut out from the thin-walled section (see Fig. 4.12) is

$$\frac{\partial t}{\partial \zeta} = -\delta(\zeta) \frac{\partial \sigma}{\partial x}\,, \tag{4.57}$$

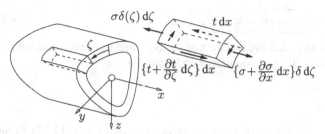

Fig. 4.12 Element of a thin-walled profile

and, with (4.55)

$$\frac{\partial t}{\partial \zeta} = -E\delta(\zeta)\psi(\zeta)\vartheta''(x).$$ (4.58)

4.4.1 Closed profiles

The resulting torque is (4.34)

$$M_T = \oint a(\zeta)t(x,\zeta)\,d\zeta,$$ (4.59)

where now $t(x,\zeta)$ is no longer constant. Differentiating function $\psi(\zeta)$ (Eq. 4.44) with respect to ζ yields

$$a(\zeta) = -\frac{d\psi}{d\zeta} + \frac{2A_m}{\delta(\zeta)}\left\{\oint \frac{d\zeta}{\delta(\zeta)}\right\}^{-1}.$$ (4.60)

Introducing this result into (4.59) leads to

$$M_T = -\oint \frac{d\psi}{d\zeta}t(x,\zeta)\,d\zeta + GJ_T\vartheta(x),$$ (4.61)

where use has been made of Eqs. (4.35), (4.41), and (4.42)$_2$. Integration by parts of the first integral then allows the introduction of Eq. (4.58), and thus, we finally arrive at

$$M_T = GJ_T\vartheta(x) - EC_T\vartheta''(x) = M_T^* + M_T^{**}$$ (4.62)

with

$$C_T = \oint \psi^2(\zeta)\delta(\zeta)\,d\zeta,$$ (4.63)

the warping resistance.

The differential equation

$$\vartheta'' - \lambda^2\vartheta = -\frac{M_T}{EC_T}, \quad \lambda^2 = \frac{GJ_T}{EC_T},$$ (4.64)

has the general solution

$$\boxed{\vartheta(x) = c_1 \sinh \lambda x + c_2 \cosh \lambda x + \vartheta_p\,,} \qquad (4.65)$$

where ϑ_p is the particular solution, and the constants c_1 and c_2 are determined from the boundary conditions. Typical boundary conditions are, e.g.

$$\vartheta = 0 \quad \text{fixed end}$$
$$\vartheta' = 0 \quad \text{free end.}$$

When $\vartheta(x)$ is known, the stress components are determined by Eqs. (4.55), (4.58), and the formulas of the St. Venant theory.

We realize that the torque M_T in Eq. (4.62) has been split into two parts, where the first term describes the problem of pure torsion with free warping (St. Venant's theory) whereas the second term is related with the restraining of warping.

4.4.2 Open profiles

This is also true for open profiles,

$$M_T = M_T^* + M_T^{**}, \qquad (4.66)$$

where now

$$M_T^* = GJ_T\vartheta(x) \qquad (4.67)$$

is related with a vanishing shear flow $t^* = 0$, and J_T is given by Eq. (4.46). The second term, as before, is necessary to prevent the cross section from warping,

$$M_T^{**} = \int_L a(\zeta)t(x,\zeta)\,\mathrm{d}\zeta\,. \qquad (4.68)$$

We note again that here the distance $a(\zeta)$ is measured from the elastic line.

In a similar procedure, we can show that the above differential equation (4.64) is also valid for thin-walled open sections, where J_T is given by Eq. (4.46), and the warping resistance may be calculated from

$$C_T = \int_L \psi^2(\zeta)\delta(\zeta)\,\mathrm{d}\zeta\,. \qquad (4.69)$$

Example 4.5:

A cantilever beam is subject to a torque M_T. Warping is prevented at one end ($x = 0$). Determine the twist ϑ and the axial stresses σ_{xx}.

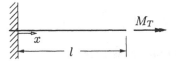

Solution:

At $x = l$ a torque M_T is applied, and the axial stress σ_{xx} is zero. Here M_T is constant, and the general solution (Eq. 4.65) becomes

$$\vartheta = \frac{M_T}{GJ_T}(1 + c_1 \sinh \lambda x + c_2 \cosh \lambda x), \quad \lambda^2 = \frac{GJ_T}{EC_T}.$$

The end conditions are $u_x = 0$ at $x = 0$, and $\sigma_{xx} = 0$ at $x = l$. By Eq. (4.56), and the equation $u_x = \vartheta\psi$, these require

$$\vartheta(0) = 0 \quad \text{and} \quad \vartheta'(l) = 0,$$

giving

$$c_1 = \tanh \lambda l, \quad c_2 = -1.$$

The rotation of cross section is

$$\varphi = \int_0^x \vartheta \, dx = \frac{M_T}{GJ_T}\left\{x + \frac{1}{\lambda}\tanh \lambda l(\cosh \lambda x - 1) - \frac{1}{\lambda}\sinh \lambda x\right\}.$$

Then, from Eq. (4.55), we find

$$\sigma_{xx}(x, \zeta) = E\psi\vartheta' = \frac{M_T}{GJ_T}\lambda E\psi(\zeta)\{\tanh \lambda l \cosh \lambda x - \sinh \lambda x\}.$$

Example 4.6:

A cantilever beam is subject to a uniformly distributed torsional load m_T. Warping is prevented at $x = 0$. Determine the twist ϑ and the axial stresses σ_{xx}.

$m_T = 15$ kNm/m, $\quad l = 10$ m.

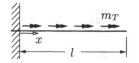

Solution:

The load m_T is per unit length, and the torque at point x is

$$M_T = m_T(l - x).$$

The end conditions are $u_x = 0$ at $x = 0$, and $\sigma_{xx} = 0$ at $x = l$, giving

$$\vartheta(0) = 0 \quad \text{and} \quad \vartheta'(l) = 0.$$

The solution of Eq. (4.65) satisfying these conditions is

$$\vartheta = \frac{m_T l}{GJ_T}\left(1 - \frac{x}{l} + c_1 \sinh \lambda x + c_2 \cosh \lambda x\right),$$

with

$$c_1 = \frac{1 + \lambda l \sinh \lambda l}{\lambda l \cosh \lambda l}, \quad c_2 = -1.$$

From Eq. (4.55), we find

$$\sigma_{xx}(x, \zeta) = \frac{m_T}{GJ_T}E\psi(\zeta)\{-1 + \lambda l c_1 \cosh \lambda x - \lambda l \sinh \lambda x\}.$$

and the rotation of cross sections is

$$\varphi = \int_0^x \vartheta \, dx = \frac{m_T l^2}{GJ_T} \left\{ \frac{x}{l} - \frac{1}{2}\left(\frac{x}{l}\right)^2 - \frac{1}{\lambda}\tanh \lambda l (\cosh \lambda x - 1) - \frac{1}{\lambda}\sinh \lambda x \right\}.$$

Introducing now the cross-sectional quantities of Examples 3.4 and 4.2, we first determine the magnitude of λ. From Eq. (4.63), the warping resistance of a closed cross section is

$$C_T = \oint \psi^2(\zeta)\delta(\zeta) \, d\zeta \,.$$

Thus using the distribution of the warping function, we obtain after integration

$$C_T = 19.286 \cdot 10^6 \text{ cm}^6,$$

and from Eq. (4.64)₂

$$\lambda^2 = \frac{J_T}{2(1+\nu)C_T} = 1.154 \cdot 10^{-2} \text{ cm}^{-2}$$
$$\lambda = 1.074 \cdot 10^{-1} \text{ cm}^{-1} \quad \rightarrow \quad \lambda l = 1.074 \cdot 10^2,$$

where Poisson's ratio has been introduced as $\nu = 0.3$. With this information, we are able to describe the behaviour of the cantilever beam. The maximum value of the

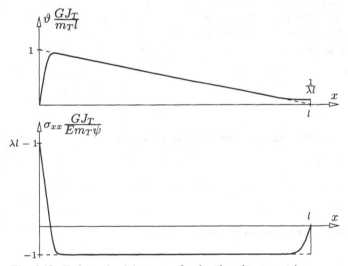

Fig. 4.13 Twist and axial stresses for the closed cross section

normal stress is obtained at $x = 0$

$$\sigma_{xx}(0) = \frac{m_T}{GJ_T} E\psi(\zeta)(\lambda l - 1)|_{\zeta=0} = \pm 23.05 \text{ MPa}$$

with the maximum value of the warping function $\psi(\zeta)$ at $\zeta = 0$ (see Example 4.2).

These values are now compared with the results for the open cross section. Using the warping function $\psi(\zeta)$, we obtain

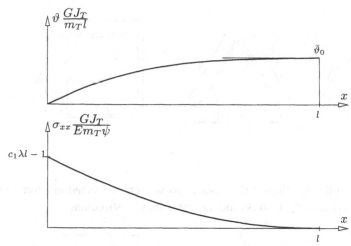

Fig. 4.14 Twist and axial stresses for the open cross section

$$C_T = 2\,723.21 \cdot 10^6 \text{ cm}^6,$$

and thus from Eq. $(4.64)_2$

$$\lambda l = 32.47 \cdot 10^{-2},$$

which is by a factor of 330.7 smaller than that for the closed cross section.

The maximum value of the normal stress is obtained at $x = 0$

$$\sigma_{xx}(0) = \pm 1\,587.16 \text{ MPa},$$

which would be far too high in a real structure, and with the maximum value of the warping function $\psi(\zeta)$ at $\zeta = 0$ (see Example 4.2). The twist attains its maximum value asymptotically at $x = l$

$$\bar{\vartheta}_0 = \vartheta \frac{GJ_T}{m_T l}\bigg|_{x=l} = \frac{1}{\lambda l}\left[\tanh \lambda l - \frac{\lambda l}{\cosh \lambda l}\right].$$

The difference between the two solutions for the closed (Fig. 4.13) and the open cross section (Fig. 4.14), respectively, is mainly based on the different magnitudes of the length scale parameter λl. For larger values of λl the particular solution of the differential equation (4.65) (see the dashed lines in Fig. 4.13) becomes predominant, with deviations according to the boundary conditions only in the small regimes close to the boundaries. With decreasing λl, these 'disturbances' of the particular solution increase until the whole solution is predominated by the boundary conditions.

4.5 Exercises to Chapter 4

Problem 4.1:

Each of the following thin-walled closed cross sections is subjected to a torque M. Compute the maximum shear stress and the shear stress distribution.

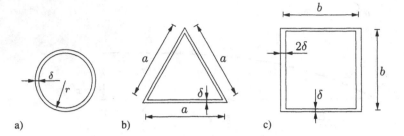

Problem 4.2:

For each of the following thin-walled open cross sections subjected to a torque M, compute the maximum shear stress and the shear stress distribution.

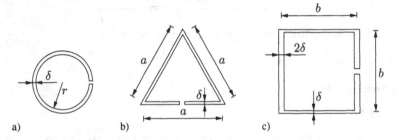

Problem 4.3:

A thin-walled closed cross section is subjected to a shear force Q as shown in the diagram. Compute the maximum shear stress and the shear stress distribution.

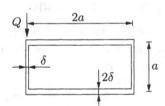

Problem 4.4:

The thin-walled open cross section is subjected to a shear force Q and a torque M as shown in the diagram. For $a = 20$ cm, $\delta = 1.5$ cm, $Q = 400$ kN, and $M = 1.8$ kNm, compute

1. the shear stress distribution across the cross section due to the shear force Q,
2. and the total shear stress distribution across the thickness at the point b-b shown in the diagram.

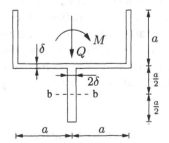

Problem 4.5:

A circular tube and a square tube are built in at one end and subjected to the same torque at the other end. Both tubes have the same length, same wall thickness, same cross-sectional area, and are constructed of the same material. What are the ratios of their shear stresses and angle of twist? (Disregard the effects of stress concentrations at the corner of the square tube.)

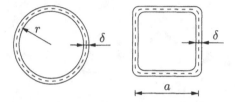

Problem 4.6:

A cantilever beam of length l is subjected to a torque M at its free end A. For each of the cross sections shown in the figure, find the shear stress distribution and the angle of twist at Point A. The cross section a) is a circle with diameter a, b) is a square where the length of each side is a, c) is an equilateral triangle where each side has a length of a, and d) is a rectangle with dimensions $a \times a/2$.

$l = 4$ m, $a = 30$ cm, $M_T = 100$ kNm, $G = 8.1 \cdot 10^4$ N/mm^2.

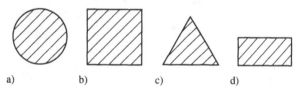

a)   b)   c)   d)

5. Curved Beams

5.1 General, Statics

Development of the theory of bending of beams was based on the assumption that, before deformation, all fibers of a small beam element have the same length $\mathrm{d}x$. If the beam is initially straight, this assumption is true, and it can serve as a good approximation if the curvature of the beam is small. However, for beams with considerable initial curvature, this curvature has to be taken into account.

From the statics of curved beams, we know the differential equations for the resultants $N(s)$, $Q(s)$ and $M(s)$

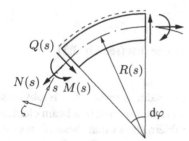

Fig. 5.1
Beam element

$$\frac{\mathrm{d}N(s)}{\mathrm{d}s} = -\frac{Q(s)}{R(s)} - n(s),$$

$$\frac{\mathrm{d}Q(s)}{\mathrm{d}s} = \frac{N(s)}{R(s)} - q(s), \qquad (5.1)$$

$$\frac{\mathrm{d}M(s)}{\mathrm{d}s} = Q(s) - m(s),$$

where a new coordinate s has been introduced in the direction of the curved beam axis, such that

$$x = x(s), \quad z = z(s), \qquad (5.2)$$

and $n(s)$, $q(s)$ and $m(s)$ are the distributed forces and moments, respectively, acting on the curved beam axis. As shown in Fig. 5.1, $R(s)$ denotes the radius of curvature at any point s of the beam axis. Moreover, with the transformation

$$\mathrm{d}s = R(s)\,\mathrm{d}\varphi, \qquad (5.3)$$

we may introduce an alternative description of Eqs. (5.1) as functions of angle φ, which sometimes is of advantage. We note that in the limit for vanishing curvature $1/R$ these differential equations take the same form as Eqs. (3.43), where then s is replaced by x.

It will be assumed here that the axis of the beam is a plane curve, and that the beam has a plane of symmetry. It is further assumed that, at the cross section considered, the following stress resultants are acting:

$$N = N(s), \quad Q = Q(s), \quad M = M(s). \tag{5.4}$$

In addition to these loads, the increase of a temperature field Θ is also taken into consideration.

To describe the stresses and strains at the cross section, and the displacements of the beam axis, a second coordinate ζ is introduced perpendicular to s and directed as the radii of the beam, such that the deflections u and w are the components of the displacement vector of the beam axis in the s and ζ directions, respectively.

The elementary theory of curved beams is again based on the assumption that plane cross sections remain plane during deformation. Possible effects of transverse shearing forces on the deformations of the beam are thus disregarded (Bernoulli's hypothesis).

5.2 Large Curvature

As in the theory of straight beams, we further assume that (see Eq. 3.3)

$$\sigma_{yy} = \sigma_{\zeta\zeta} = \sigma_{y\zeta} = 0. \tag{5.5}$$

We note that the assumption of disregarding the radial stresses $\sigma_{\zeta\zeta}$ is obviously inconsistent with the condition of equilibrium in radial direction of a beam element. However, a two-dimensional analysis of this problem shows that these stresses are considerably smaller than the longitudinal stresses $\sigma_{ss} = \sigma(s, \zeta)$.

In the course of bending, the cross section will rotate with respect to its original position by the angle $\Delta\,d\varphi$, and the different fibers undergo different exten-

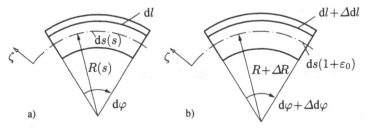

Fig. 5.2 Undeformed (a) and deformed (b) beam element

sions/compressions $\Delta\,dl$, relative to their distance ζ from the beam axis

$$\Delta\,dl = \varepsilon_0(s)\,ds + \zeta\Delta\,d\varphi, \tag{5.6}$$

where

$$\varepsilon_0(s) = \frac{\Delta \, ds}{ds} = \frac{du(s)}{ds} + \frac{w(s)}{R(s)} \tag{5.7}$$

is the extensional strain of the beam axis. Thus the strain $\varepsilon_{ss} = \varepsilon(s, \zeta)$ in a fiber located a distance ζ from the axis is

$$\varepsilon(s, \zeta) = \frac{\Delta \, dl}{dl} = \frac{\varepsilon_0(s) + \zeta \dfrac{\Delta \, d\varphi}{ds}}{1 + \zeta \dfrac{d\varphi}{ds}}, \tag{5.8}$$

and the change of curvature of the beam axis is

$$\begin{aligned}
\kappa(s) &= \frac{1}{R + \Delta R} \cdot \frac{1}{R} = \frac{1}{1 + \varepsilon_0(s)} \left\{ \frac{\Delta \, d\varphi}{ds} - \frac{\varepsilon_0(s)}{R} \right\} \\
&= \frac{\Delta \, d\varphi}{ds} - \frac{\varepsilon_0(s)}{R} = -\frac{d^2 w(s)}{ds^2} + \frac{d}{ds}\left(\frac{u(s)}{R(s)} \right) - \frac{\varepsilon_0(s)}{R},
\end{aligned} \tag{5.9}$$

where (since $\varepsilon_0 \ll 1$) ε_0 is neglected against unity, and

$$dl = ds + \zeta \, d\varphi, \quad d\varphi = \frac{ds}{R(s)}. \tag{5.10}$$

Moreover, the small angle $\Delta \, d\varphi$ is described by the small change of the slope of the beam axis

$$\Delta \, d\varphi = -d\left(\frac{dw}{ds} \right) + d\left(\frac{u}{R} \right). \tag{5.11}$$

Introducing these results into Eq. (5.8), we finally arrive at

$$\varepsilon(s, \zeta) = \varepsilon_0(s) + \frac{R \kappa(s)}{R + \zeta} \zeta, \tag{5.12}$$

which is comparable with Eq. (3.5) for the straight beam. In accordance with Hooke's law, the corresponding normal stress is then

$$\sigma(s, \zeta) = E \left\{ \varepsilon_0(s) + \frac{R \kappa(s)}{R + \zeta} \zeta \right\}. \tag{5.13}$$

The stress resultants N and M are defined as

$$N(s) = \int_A \sigma(s, \zeta) \, dA, \quad M(s) = \int_A \zeta \sigma(s, \zeta) \, dA. \tag{5.14}$$

Thus we find

$$\boxed{\begin{aligned}
N(s) &= E \left\{ \varepsilon_0(s) A(s) - \kappa(s) \frac{J^*(s)}{R(s)} \right\}, \\
M(s) &= \kappa(s) E J^*(s),
\end{aligned}} \tag{5.15}$$

with the usual definition of the area, and a modified moment of inertia

$$J^*(s) = R \int_A \frac{\zeta^2}{R+\zeta} \, dA , \tag{5.16}$$

and where we further have made use of the fact that the first moment of the area of the cross section with respect to the y axis vanishes for a system of principal axes.

From Eqs. (5.15), we determine

$$
\begin{aligned}
\varepsilon_0(s) &= \frac{N(s)}{EA(s)} + \frac{M(s)}{EA(s)\,R(s)} , \\
\kappa(s) &= \frac{M(s)}{EJ^*(s)} .
\end{aligned}
\tag{5.17}
$$

and finally

$$\boxed{\sigma(s,\zeta) = \frac{N(s)}{A(s)} + \frac{M(s)}{J^*(s)} \left\{ \frac{J^*(s)}{A(s)R(s)} + \frac{R(s)\zeta}{R(s)+\zeta} \right\} .} \tag{5.18}$$

The distribution of the normal stress over the depth of the cross section is no longer linear, but follows a hyperbolic law; across the width of the cross section, they are distributed uniformly.

From the condition $\sigma = 0$, the neutral axis can be determined, which now no longer passes through the centroid of the cross section; if $N = 0$,

$$\boxed{\zeta_0 = - \frac{R(s)J^*(s)}{J^*(s) + A(s)R^2(s)} .} \tag{5.19}$$

If the normal stresses are known, the shear stresses may be calculated in the usual manner (see Section 3.2). For a rectangular cross section we get

$$\boxed{\tau(s,\zeta) = \frac{Q(s)S(\zeta)}{b\,J^*} \left\{ \frac{R}{R+\zeta} \right\}^2 .} \tag{5.20}$$

Example 5.1:
Determine the normal stresses in a crane hook subject to a vertical load F. The critical section may be approximated by a rectangular cross section b/h, with $R_i = h/2$, the inner radius of the hook.

Solution:
We first calculate the resultants acting in the beam axis

$$N = F, \quad M = -FR,$$

where $R = R_i + h/2 = h$, and $R_a = R_i + h$. The modified moment of inertia is

$$J^* = R \int_{-h/2}^{h/2} \frac{\zeta^2 b}{R+\zeta} \, d\zeta = R^3 b \left\{ \ln \frac{R_a}{R_i} - \frac{h}{R} \right\} = bh^3(\ln 3 - 1) .$$

Introducing this into Eq. (5.18), we find

$$\sigma = \frac{F}{A} - \frac{FR}{J^*}\left\{\frac{J^*}{AR} + \frac{R\zeta}{R+\zeta}\right\} = -\frac{F}{A(\ln 3 - 1)}\frac{\zeta}{h+\zeta}$$

$$= \frac{F}{A} \cdot \begin{cases} 10,14 \text{ inner radius} \quad (7) \\ -3,38 \text{ outer radius} \ (-5) \end{cases}.$$

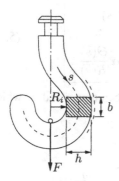

Fig. 5.3
Stresses in a crane hook

The numbers in the brackets designate the respective factors which would be obtained from the straight beam theory.

5.3 Small Curvature

If the depth of the beam is small compared to the radius of initial curvature, then ζ can be neglected against R in the foregoing equations. The neutral axis passes through the centroid of the cross section (if $N = 0$), and, hence, for the normal and shear stresses, one obtains expressions which are identical with those for straight beams. Thus, in curved beams having a small depth, the stresses can be calculated as though the beam were straight.

5.4 Deflections of Curved Beams

In the same manner as for straight beams, we will utilize the relations between the change of curvature $\kappa(s)$ and the strain of the beam axis $\varepsilon_0(s)$, and the displacements u and w to develop the differential equations for these displacements. From Eqs. (5.7) and (5.9) we find

$$\varepsilon_0(s) = u'(s) + \frac{w(s)}{R(s)} - \alpha\Theta, \tag{5.21}$$

and

$$\kappa(s) = -w''(s) + \left(\frac{u}{R}\right)' - \frac{\varepsilon_0(s)}{R}, \tag{5.22}$$

where now the prime designates differentiation with respect to s, and where moreover the influence of a uniform heating is taken into consideration. Introducing Eq. (5.21) into Eq. (5.22), we get

$$\kappa(s) = -w''(s) - \frac{w(s)}{R^2(s)} + u\left(\frac{1}{R}\right)' + \alpha\frac{\Theta}{R}, \qquad (5.23)$$

and finally with Eqs. (5.14)

$$
\begin{array}{|l|}
\hline
u'(s) + \dfrac{w(s)}{R(s)} = \dfrac{1}{EA(s)}\left\{N(s) + \dfrac{M(s)}{R(s)}\right\} + \alpha\Theta(s) \\[3mm]
w''(s) + \dfrac{w(s)}{R^2(s)} - u(s)\left(\dfrac{1}{R(s)}\right)' = -\dfrac{M(s)}{EJ^*(s)} + \alpha\dfrac{\Theta(s)}{R(s)} \, . \\
\hline
\end{array}
\qquad (5.24)
$$

These are two coupled differential equations for the displacements u and w, which can be decoupled for circular arches, i.e., if $R = \text{const}$.

Example 5.2:

For the clamped circular arch subjected to a constant radial load q as shown in the figure, compute the vertical and horizontal deflections of point A.

Solution:

The system is statically determinate. We first determine the reactions at the clamp to be

$$
\begin{aligned}
H &= -Q(0) = qR, \\
V &= -N(0) = qR, \\
M &= M(0) = qR^2.
\end{aligned}
$$

From Eqs. (5.1), we find the differential equation

$$Q''(\varphi) + Q(\varphi) = 0,$$

where we have made use of Eq. (5.3), and where here the prime stands for differentiation with respect to the angle φ. The solution of this equation is

$$Q(\varphi) = c_1 \sin\varphi + c_2 \cos\varphi.$$

With this solution, we find for the additional resultants

$$
\begin{aligned}
N(\varphi) &= Q'(\varphi) - qR = c_1 \cos\varphi - c_2 \sin\varphi - qR, \\
M(\varphi) &= \int Q(\varphi)R\,d\varphi = -c_1 R\cos\varphi + c_2 R\sin\varphi + c_3 \, .
\end{aligned}
$$

With the above initial conditions, we arrive at

$$c_1 = 0, \quad c_2 = -qR, \quad c_3 = qR^2,$$

and thus

$$N(\varphi) = -qR(1 - \sin\varphi),$$
$$M(\varphi) = qR^2(1 - \sin\varphi).$$

Introducing this result into Eqs. (5.24), the differential equations for the deflections read

$$u'(\varphi) + w(\varphi) = \frac{1}{EA}\{M(\varphi) + RN(\varphi)\} = 0$$

$$w''(\varphi) + w(\varphi) = -\frac{M(\varphi)}{EJ}R^2 = -\frac{qR^4}{EJ}(1 - \sin\varphi).$$

where we have assumed the curvature of the beam to be small.

The solution of the second differential equation consists of two parts

$$w(\varphi) = w_h(\varphi) + w_p(\varphi),$$

where

$$w_h(\varphi) = c_4 \sin\varphi + c_5 \cos\varphi$$

is the solution of the homogeneous equation, and

$$w_p(\varphi) = -\frac{qR^4}{2EJ}(2 + \varphi \cos\varphi)$$

is a special solution of the non-homogeneous equation. Thus we find

$$w(\varphi) = c_4 \sin\varphi + c_5 \cos\varphi - \frac{qR^4}{2EJ}(2 + \varphi \cos\varphi).$$

The constants c_4, c_5 may be determined from the kinematical conditions

$$w(0) = w'(0) = 0$$

as

$$c_4 = \frac{qR^4}{2EJ}, \quad c_5 = \frac{qR^4}{EJ}.$$

With these informations we finally arrive at

$$w(\varphi) = \frac{qR^4}{2EJ}(\sin\varphi + 2\cos\varphi - \varphi\cos\varphi - 2).$$

Now from the first differential equation, we compute

$$u(\varphi) = -\int w(\varphi)\,d\varphi = \frac{qR^4}{2EJ}(2\cos\varphi - 2\sin\varphi + 2\varphi + \varphi\sin\varphi) + c_6.$$

Again, from the kinematical condition $u(0) = 0$ follows

$$c_6 = -\frac{qR^4}{EJ},$$

and

$$u(\varphi) = \frac{qR^4}{2EJ}\,(2\cos\varphi - 2\sin\varphi + 2\varphi + \varphi\sin\varphi - 2).$$

Thus the deflections of point A are:

$$EJw_A = -\frac{1}{2}qR^4, \quad EJu_A = \left(\frac{3}{4}\pi - 2\right)qR^4.$$

5.5 Exercises to Chapter 5

Problem 5.1:

The semi-circular beams have flexural rigidity EJ, axial rigidity EA and are loaded as shown in the diagrams. Determine in each case for the point of application of the load the vertical and horizontal displacements, the curve $u(\varphi)$ for the displacement in tangential direction and the equation $w(\varphi)$ for the displacement in radial direction. Assume in each case that the curvature of the beam is small.

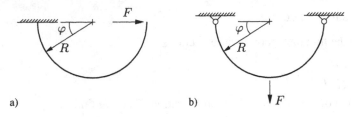

a)   b) F

Problem 5.2:

The circular beams have flexural rigidity EJ, axial rigidity EA and are loaded as shown in the diagrams. Determine in each case for the point of application of the load the vertical and horizontal displacements, the curve $u(\varphi)$ for the displacement in tangential direction and the equation $w(\varphi)$ for the displacement in radial direction. Assume in each case that the curvature of the beam is small.

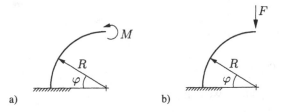

a)   b)

6. Simple Beams II: Energy Principles

6.1 Reciprocity Theorems of Betti and Maxwell

Consider two groups of loads applied to an elastic body. The loads $\{b_1, s_1\}$ produce displacements u_1, and the loads $\{b_2, s_2\}$ produce displacements u_2. The law of superposition states that the simultaneous application of both load groups produces the displacements $u_1 + u_2$. This law is based on the linearity of the fundamental equations, used so far.

If $\{b, s\}$ are the final values of the external forces, and if u are the corresponding displacements, then the work done during loading is

$$A = A^* = \frac{1}{2} \int_V b \cdot u \, dV + \frac{1}{2} \int_A s \cdot u \, dA. \tag{6.1}$$

As in Eqs. (2.18) and (2.20), the factor $\frac{1}{2}$ is a consequence of the proportionality of load and displacement during loading. Equation (6.1) may be written symbolically as

$$A = \frac{1}{2} F \cdot u, \tag{6.2}$$

where as in Section 2.7, F are generalized forces, and u are generalized displacements.

We now consider two load systems, F_1 and F_2. While F_1 is applied, the work

$$A_{11} = \frac{1}{2} F_1 \cdot u_1 = \frac{1}{2} F_1 f_{11} = \frac{1}{2} \delta_{11} F_1^2 \tag{6.3}$$

is done, where $f_{11} = \delta_{11} F_1$ is the component of the displacement field u_1 at point 1 and in the direction of F_1, and δ_{11} is a so called influence coefficient, expressing the proportionality between load and displacement.

When F_2 is added, work of two kinds is done. The forces F_2 do the work

$$A_{22} = \frac{1}{2} F_2 \cdot u_2 = \frac{1}{2} F_2 f_{22} = \frac{1}{2} \delta_{22} F_2^2 \tag{6.4}$$

with $f_{22} = \delta_{22} F_2$, the component of u_2 at point 2 and in the direction of F_2, and the forces F_1 do the work

$$A_{12} = F_1 \cdot u_2 = F_1 f_{12} = F_1 \delta_{12} F_2 \tag{6.5}$$

with $f_{12} = \delta_{12} F_2$, the component of u_2 at point 1 and in the direction of F_1.

The total work done is

$$A = A_{11} + A_{12} + A_{22} .$$ (6.6)

Because of the superposition law, this work must be independent of the order in which the loads have been applied, i.e. it must be equal to

$$A = A_{22} + A_{21} + A_{11} .$$ (6.7)

with

$$A_{21} = \mathbf{F}_2 \cdot \mathbf{u}_1 = F_2 f_{21} = F_2 \delta_{21} F_1 ,$$ (6.8)

and, therefore

$$\boxed{A_{12} = A_{21} .}$$ (6.9)

This is Betti's theorem: When two systems of forces act on an elastic body, the work of the forces 1 on the displacements 2 is equal to the work of the forces 2 on the displacements 1.

Maxwell's reciprocity theorem is a special case of Betti's theorem. We define the influence coefficient δ_{ik} as the displacement at point i resulting from a unit force at the point k. Then Maxwell's reciprocity theorem states

$$\boxed{\delta_{ik} = \delta_{ki} .}$$ (6.10)

Note that herein the forces as well as the displacements are used in a generalized sense.

6.2 Theorems of Engesser and Castigliano

Consider a linear elastic body subject to several loads $\mathbf{F}_i$ $(i = 1, 2, \ldots, n)$. The points where these loads are acting as well as their directions are considered known. Then the influence coefficients δ_{ik} are fixed. Only the scalar magnitudes of the F_i are varying.

The total work done on this body is

$$A = \frac{1}{2} \sum_i \sum_k F_i \delta_{ik} F_k = A(F_i), \quad i, k = 1, 2, \ldots, n .$$ (6.11)

Differentiating A with respect to a specific force F_i,

$$\frac{\partial A}{\partial F_i} = \sum_k \delta_{ik} F_k = \sum_k f_{ik} = f_i ,$$ (6.12)

gives the displacement of point i in the direction of F_i. Moreover, a second differentiation of this expression with respect to the force F_k,

$$\frac{\partial^2 A}{\partial F_i \partial F_k} = \delta_{ik},$$ (6.13)

describes the influence coefficients δ_{ik}. Since $A = W^*$ for a linear elastic system, we arrive at (Eq. 2.60)

$$\boxed{\frac{\partial W^*(F_i)}{\partial F_k} = f_k\,,}$$

(6.14)

Engesser's theorem. Expressing now $W = W(f_i)$, from Section 1.6 and the definition

$$dW = \sum_i F_i\,df_i = \sum_i \frac{\partial W}{\partial f_i}\,df_i\,,$$

(6.15)

we find

$$\boxed{\frac{\partial W(f_i)}{\partial f_k} = F_k\,,}$$

(6.16)

Castigliano's second theorem.

6.3 Statically Indeterminate Systems

A structure is statically indeterminate if the total number of forces and moments, both internal and external, is greater than the number which can be calculated from equations of static equilibrium. The degree of indeterminacy is the number of these quantities in excess of the number of equilibrium equations. The excess forces or reactions may be called redundants, when some or all of the redundants are removed, the structure which remains is called the primary structure or released structure. Usually, the primary structure is selected so as to be statically determinate, but this is not always necessary.

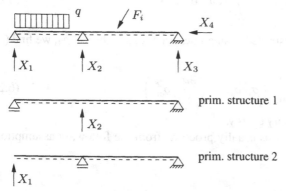

Fig. 6.1 Statically indeterminate system of degree 1: different primary structures

To solve statically indeterminate systems, we first may introduce the removed redundants as external loads X_k on the primary structure, and thus get

$$W^* = W^*(F_i, X_k).$$ (6.17)

We also can select internal resultants as redundants, e.g. the internal bending moment at the central support. Then the primary structure has no restraint against bending moment; this is obtained by inserting a hinge into the beam.

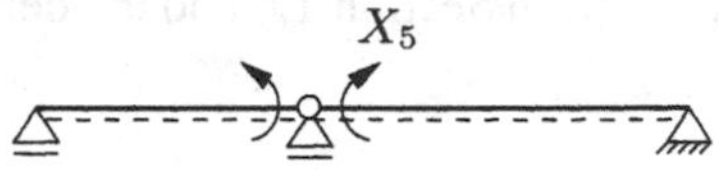

X_5

Fig. 6.2
Primary structure 3

In all these cases, in the real statically indeterminate structure, we have $f_k = 0$, and thus from Eq. (6.14) follows

$$\boxed{\frac{\partial W^*(F_i, X_k)}{\partial X_k} = 0,}$$ (6.18)

Menabrea's theorem.

Again, this is a stationarity condition for the complementary energy as function of the unknown redundants X_k. Since (see Eq. 6.13)

$$\frac{\partial^2 W^*}{\partial X_k^2} = \delta_{kk} > 0,$$ (6.19)

this proves that the variational principle (2.57) is a minimum principle.

6.4 The Complementary Energy of Beams

In Section 2.2, we introduced the strain and complementary energies of deformable bodies,

$$dW = \int_V \rho\, dw\, dV, \quad dW^* = \int_V \rho\, dw^*\, dV.$$ (6.20)

As our considerations are restricted to small deformations with $\rho \cong \rho_0$, we find (see e.g. Eq. 2.21)

$$\rho w^* = \int \varepsilon_{ik}\, d\sigma_{ik} = \frac{1}{4G}\left(\sigma_{ik}\sigma_{ik} - \frac{9\nu}{1+\nu}\sigma_m^2\right)$$ (6.21)

for the specific complementary energy.

The theory of simple beams usually proceeds from the following assumptions (see Eq. 3.3):

$$\sigma_{yy} = \sigma_{zz} = \sigma_{yz} = 0.$$ (6.22)

Thus the remaining stress components acting in the cross section of a beam are:

σ_{xx} — normal stresses, and

$\tau = \sqrt{\sigma_{xy}^2 + \sigma_{xz}^2}$ — shear stresses, (6.23) where according to different cross sections different descriptions for the shear stresses are used (refer to Section 3.2).

This leads us to

$$\rho w^* = \frac{1}{2E} \sigma_{xx}^2 + \frac{1}{2G} \tau^2 . \tag{6.24}$$

in a linear theory, and

$$W^* = \int_L \int_A \left\{ \frac{\sigma_{xx}^2}{2E} + \frac{\tau^2}{2G} \right\} \mathrm{d}A \, \mathrm{d}l . \tag{6.25}$$

For straight beams, all fibers of a small element have the same length. Thus we can introduce

$$\mathrm{d}l = \mathrm{d}x . \tag{6.26}$$

This assumption can also serve as a good approximation if the curvature of the beam is small. In this case, we simply replace $\mathrm{d}x$ by $\mathrm{d}s$. However, for beams with considerable initial curvature, the different lengths of fibers at different positions of the cross section cannot be disregarded. In this case Eq. $(5.10)_1$ has to be taken into account.

For the moment, we restrict our considerations to straight beams, and beams having small curvatures. Then the different loads acting on the beam will contribute to the complementary energy W^* of the beam. In a linear theory these contributions are from different sources:

(i)   extension,
(ii) bending in two planes,
(iii) torsion, and
(iv) shear deformations.

Due to the principle of superposition, these contributions may be simply added. We note, however, that shear deformations during bending are excluded by Bernoulli's hypothesis, and thus the appearance of shear stresses due to these shear deformations is inconsistent with Hooke's law (refer to Section 3.2).

The normal stresses σ_{xx} caused by extension and bending in two planes can be expressed by using Eq. (3.16)

$$\sigma_{xx} = \frac{N}{A} + \frac{M_y(x)}{J_{yy}} z - \frac{M_z(x)}{J_{zz}} y . \tag{6.27}$$

Introducing this into Eq. (6.25), the first contribution due to the normal stress to the total complementary energy becomes

$$W_1^* = \int_L \frac{1}{2E} \left\{ \frac{N^2}{A^2} \int_A \mathrm{d}A + \frac{M_y^2}{J_{yy}^2} \int_A z^2 \, \mathrm{d}A + \frac{M_z^2}{J_{zz}^2} \int_A y^2 \, \mathrm{d}A \right.$$
$$\left. +2 \frac{N}{A} \frac{M_y}{J_{yy}} \int_A z \, \mathrm{d}A - 2 \frac{N}{A} \frac{M_z}{J_{zz}} \int_A y \, \mathrm{d}A - 2 \frac{M_y}{J_{yy}} \frac{M_z}{J_{zz}} \int_A zy \, \mathrm{d}A \right\} \mathrm{d}x . \tag{6.28}$$

If the y and z axes are the centroidal principal axes of inertia, then the last three terms in the foregoing integral vanish, and the complementary energy due to extension and bending is

$$W_1^* = \frac{1}{2E} \int_L \left\{ \frac{N^2}{A} + \frac{M_y^2}{J_{yy}} + \frac{M_z^2}{J_{zz}} \right\} dx \,. \tag{6.29}$$

The complementary energy in a beam due to the normal stresses can thus be obtained by calculating separately and then adding the complementary energy produced by the normal-force component and the complementary energy produced by the two bending-moment components. This superposition of portions of complementary energy is possible because the normal force in a straight beam does no work on displacements produced by the bending moments, and vice versa.

The shear stresses due to torsion - without restrained warping - can be expressed using Eq. (4.13) as

$$\tau = \sqrt{\sigma_{xy}^2 + \sigma_{xz}^2} = 2G\vartheta \sqrt{(T_{,z})^2 + (T_{,y})^2} \,, \tag{6.30}$$

where now $(\bullet)_{,x_i}$ designates the partial derivative of the quantity with respect to the coordinate x_i. Introducing this into Eq. (6.25), we get

$$W_2^* = \int_L 2G\vartheta^2 \left\{ \int_A (T_{,z})^2 \, dA + \int_A (T_{,y})^2 \, dA \right\} dx = \int_L 2G\vartheta^2 \int_A T(y,z) \, dA \, dx \,, \tag{6.31}$$

after integration by parts, and using Eq. (4.14). The twist ϑ may be expressed by (Eq. 4.19)

$$\vartheta = \frac{M_T}{GJ_T} \,, \quad J_T = 4 \int_A T(y,z) \, dA \,. \tag{6.32}$$

Thus we find

$$W_2^* = \frac{1}{2G} \int_L \frac{M_T^2}{J_T} \, dx \tag{6.33}$$

for the second contribution to the total complementary energy due to torsion, and disregarding the influence of a restrained warping.

In Section 3.2, we calculated the different shear stresses due to bending from equilibrium of the stresses at a small part cut from an element of the beam under consideration. It turned out that the different results may be expressed by (Eqs. 3.35, 3.36)

$$\sigma_{xz} = \frac{Q}{A} \kappa_{xz}(y,z) \,, \quad \sigma_{xy} = \frac{Q}{A} \kappa_{xy}(y,z) \,, \tag{6.34}$$

for plane bending of a solid cross section, and (Eq. 3.38)

$$\sigma_{x\zeta} = \tau(\zeta) = \frac{Q}{J} \frac{S(\zeta)}{\delta(\zeta)} \tag{6.35}$$

Cross section	f_s	Remarks
▣	$\frac{6}{5} = 1.20$	
●	$\frac{32}{27} = 1.185$	
○	2.0	
I	$\approx 2.0 - 2.4$	$f_s \approx \dfrac{A}{A_{\text{web}}}$
T	$\approx 3 - 4$	
⊔	$\approx 2 - 2.4$	

Table 6.1 Form factor f_s

for thin-walled cross sections. Introducing now the form factor of shear f_s as

$$f_s = \frac{1}{A} \int_A \left(\kappa_{xz}^2 + \kappa_{xy}^2 \right) \, \mathrm{d}A \tag{6.36}$$

for solid cross sections, and

$$f_s = \frac{A}{J^2} \int_A \frac{S^2(\zeta)}{\delta(\zeta)} \, \mathrm{d}\zeta \tag{6.37}$$

for thin-walled cross sections, we finally get from Eq. (6.25)

$$W_3^* = \frac{1}{2G} f_s \int_L \frac{Q^2}{A} \, \mathrm{d}x \,. \tag{6.38}$$

The form factor for shear must be evaluated for each particular shape of cross section by means of Eqs. (6.34), and (6.35) and (6.36), respectively. Some values of typical cross sections are given in Table 6.1.

The different contributions can be added to give

$$W^* = \frac{1}{2} \int_L \left\{ \frac{N^2}{EA} + \frac{M_y^2}{EJ_{yy}} + \frac{M_z^2}{EJ_{zz}} + \frac{M_T^2}{GJ_T} + f_{sz} \frac{Q_z^2}{GA} + f_{sy} \frac{Q_y^2}{GA} \right\} \mathrm{d}x. \tag{6.39}$$

Comparing the magnitudes of the different contributions to the complementary energy, we realize that whenever bending or torsion occurs in a straight member (part of a structure), the influence of the normal forces and the shear forces may be neglected. From this, we may conclude that, for practical application, Eq. (6.39) reduces to

$$W^* = \frac{1}{2} \int_L \left\{ \frac{M_y^2}{EJ_{yy}} + \frac{M_z^2}{EJ_{zz}} + \frac{M_T^2}{GJ_T} \right\} \mathrm{d}x \,, \tag{6.40}$$

and only in those members with vanishing moments and torque (and thus also vanishing shear forces), e.g. in members of trusses, is the influence of the normal forces N considered.

For beams with considerable curvature, we replace Eq. (6.26) by

$$dl = \left(1 + \frac{\zeta}{R}\right) ds,$$
(6.41)

and Eq. (6.27) by

$$\sigma = \frac{N}{A} + \frac{M}{J^*}\left\{\frac{J^*}{AR} + \frac{R\zeta}{R+\zeta}\right\},$$
(6.42)

for plane bending. Introducing these relations into Eq. (6.25), we find

$$W^* = \frac{1}{2}\int_L \left\{\frac{N^2}{EA} + 2\frac{NM}{EAR} + \frac{M^2}{EJ^*}\left[1 + \frac{J^*}{AR^2}\right]\right\} ds$$
(6.43)

having neglected here the influence of the shear deformations.

6.5 Strain Energy of Beams

In a similar way as we calculated the different contributions to the complementary energy, we may also calculate the strain energy of beams. From Eq. (2.19), we introduce the specific strain energy (where again $\rho \cong \rho_0$)

$$\rho w = \int \sigma_{ik}\, d\varepsilon_{ik} = G\left(\varepsilon_{ik}\varepsilon_{ik} + \frac{\nu}{1-2\nu}e^2\right)$$
(6.44)

and further introducing here the contributions from the different sources to the strains, we finally arrive at

$$W = \frac{1}{2}\int_L \left\{EAu'^2 + EJ_{yy}w''^2 + GJ_T\varphi'^2\right\} dx$$
(6.45)

for plane bending, and neglecting the influence of the shear deformations.

We realize that due to the linearity of the problem, since $W = W^*$, this result could also be determined from Eq. (6.39), by introducing the differential equations for the displacements u, w and φ, respectively (refer to Eqs. 3.73, here for the isothermal case, and Table 4.3).

We also note, however, that $W \neq W^*$ for the contribution of the shear deformations, since according to Bernoulli's hypothesis the latter are a priori neglected, whereas the shear stresses - calculated disregarding Hooke's law from equilibrium equations (see Section 3.2) - contribute to the complementary energy.

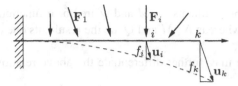

Fig. 6.3
Cantilever beam subject to different
loads

6.6 Application to Beams

Consider the cantilever beam of Fig. 6.3 subject to a group of loads containing $\mathbf{F}_i$ at point i. If the displacement f_i at point i in the direction of $\mathbf{F}_i$ is to be determined, we may apply Engesser's theorem (Eq. 6.14), and calculate

$$f_i = \frac{\partial W^*}{\partial F_i} \, . \tag{6.46}$$

This procedure, however, will not work in either of the two following situations:

(i) Determine the displacement of point i in any other direction than that of $\mathbf{F}_i$, or
(ii) determine any displacement of another point (say) k, where no load is acting.

In these cases, two strategies can be used to determine the unknown displacements:

1. We introduce at point k an additional (imaginary) force $\mathbf{F}_k$ whose direction coincides with the direction of the unknown displacement f_k. We then determine $W^* = W^*(F_i, F_k)$ and finally from Eq. (6.14)

$$\boxed{f_k = \frac{\partial W^*}{\partial F_k} \bigg|_{F_k \to 0} ,} \tag{6.47}$$

where simultaneously for the imaginary force $F_k \to 0$ is carried out.

2. We introduce a virtual unit load "1" in the direction of the unknown displacement f_k. Then from Eq. (6.39) we get, e.g.

$$f_k = \frac{\partial W^*}{\partial F_k} = \int_L \left\{ \frac{N}{EA} \frac{\partial N}{\partial F_k} + \frac{M}{EJ} \frac{\partial M}{\partial F_k} + f_s \frac{Q}{GA} \frac{\partial Q}{\partial F_k} \right\} dx , \tag{6.48}$$

for plane bending. Since, however, the resultants N, M and Q are linear functions of F_k, their partial derivatives with respect to these loads are then specific functions under unit loads $F_k = 1$, i.e.

$$\frac{\partial N}{\partial F_k} = \bar{N} , \quad \frac{\partial M}{\partial F_k} = \bar{M} , \quad \frac{\partial Q}{\partial F_k} = \bar{Q} , \tag{6.49}$$

where $\bar{N}$, $\bar{M}$ and $\bar{Q}$ are the resultants under a unit load "1" in the direction of the unknown displacement f_k. We thus arrive at

$$\boxed{f_k = \int_L \left\{ \frac{N\bar{N}}{EA} + \frac{M\bar{M}}{EJ} + f_s \frac{Q\bar{Q}}{GA} \right\} dx .} \tag{6.50}$$

We note that herein (since $F_k \to 0$) the resultants N, M and Q are linear functions of the real loads F_i, $(i = 1, \ldots, n)$, whereas $\bar{N}$, $\bar{M}$ and $\bar{Q}$ are the resultants due to virtual unit forces $F_k = 1$.

Using Eqs. (6.12) and (6.13), we may further differentiate the above relation with respect to F_i, and get

$$\delta_{ik} = \int_L \left\{ \frac{N_i N_k}{EA} + \frac{M_i M_k}{EJ} + f_s \frac{Q_i Q_k}{GA} \right\} dx \,, \tag{6.51}$$

where now N_i, M_i and Q_i are the resultants due to unit forces $F_i = 1$. This important relation makes it possible to calculate the influence coefficients δ_{ik} from product integrals of the given kind. A brief compilation of product integrals, covering the most commonly encountered functions, is given in Table 6.2.

With the help of these influence coefficients, we may reformulate Eq. (6.50) to give

$$f_k = \sum_i \delta_{ki} F_i \,. \tag{6.52}$$

Furthermore, we also may use these coefficients to solve statically indeterminate systems. For a system of degree k, we introduce the k redundants as external forces X_i on the primary structure, and thus get from Eq. (6.52)

$$0 = \delta_{k0} + \sum_i \delta_{ki} X_i \,, \tag{6.53}$$

where the δ_{k0} describe the deformations at point k of the primary structure under the given load system.

If all redundants X_i are determined from the above system of equations, we finally can describe the distributions of the resultants as, e.g. for the bending moment M,

$$M = M_0 + \sum_i M_i X_i \,, \tag{6.54}$$

where M_0 is the bending moment of the primary structure under the given load system.

Example 6.1:

The plane frame shown in the figure has a fixed support and carries a vertical load F at the free end. Both members of the frame have constant flexural rigidity EJ.
Determine the horizontal deflection δ_h, the vertical deflection δ_v, and the angle of rotation φ.

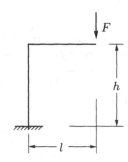

	k ⬜ k (ℓ)	triangle k (ℓ)	k_1 ⬜ k_2 (ℓ)
i ⬜ i (ℓ)	$\ell i k$	$\dfrac{1}{2}\,\ell i k$	$\dfrac{1}{2}\,\ell i(k_1 + k_2)$
triangle i (ℓ)	$\dfrac{1}{2}\,\ell i k$	$\dfrac{1}{3}\,\ell i k$	$\dfrac{1}{6}\,\ell i(k_1 + 2k_2)$
i triangle (ℓ)	$\dfrac{1}{2}\,\ell i k$	$\dfrac{1}{6}\,\ell i k$	$\dfrac{1}{6}\,\ell i(2k_1 + k_2)$
i_1 ⬜ i_2 (ℓ)	$\dfrac{1}{2}\,\ell(i_1 + i_2)k$	$\dfrac{1}{6}\,\ell(i_1 + 2i_2)k$	$\dfrac{1}{6}\,\ell(2i_1 k_1 + i_1 k_2$ $+ i_2 k_1 + 2i_2 k_2)$
quadr. parabola i (ℓ)	$\dfrac{2}{3}\,\ell i k$	$\dfrac{1}{3}\,\ell i k$	$\dfrac{1}{3}\,\ell i(k_1 + k_2)$
quadr. parabola i (ℓ)	$\dfrac{2}{3}\,\ell i k$	$\dfrac{1}{4}\,\ell i k$	$\dfrac{1}{12}\,\ell i(5k_1 + 3k_2)$
quadr. parabola i (ℓ)	$\dfrac{2}{3}\,\ell i k$	$\dfrac{5}{12}\,\ell i k$	$\dfrac{1}{12}\,\ell i(3k_1 + 5k_2)$
quadr. parabola i (ℓ)	$\dfrac{1}{3}\,\ell i k$	$\dfrac{1}{4}\,\ell i k$	$\dfrac{1}{12}\,\ell i(k_1 + 3k_2)$
quadr. parabola i (ℓ)	$\dfrac{1}{3}\,\ell i k$	$\dfrac{1}{12}\,\ell i k$	$\dfrac{1}{12}\,\ell i(3k_1 + k_2)$
i ($\alpha\ell \;\vert\; \beta\ell$, ℓ)	$\dfrac{1}{2}\,\ell i k$	$\dfrac{1}{6}\,\ell(1 + \alpha)i k$	$\dfrac{1}{6}\,\ell i\{(1 + \beta)k_1$ $+ (1 + \alpha)k_2\}$
$\displaystyle\int_0^\ell [M_k(x)]^2\,dx$	ℓk^2	$\dfrac{1}{3}\,\ell k^2$	$\dfrac{1}{3}\,\ell(k_1^2 + k_1 k_2 + k_2^2)$

Table 6.2 Product integrals $\int M_i M_k \, dx$

Solution:

The bending moment M_0 caused by the load is portrayed in Fig. 6.4, where the

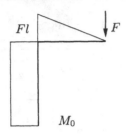

Fig. 6.4
Bending moments M_0

bending moment diagrams are plotted on the sides which are in tension. The unit loads corresponding to the horizontal deflection, vertical deflection, and angle of rotation are shown together with their bending moment diagrams in Fig. 6.5.

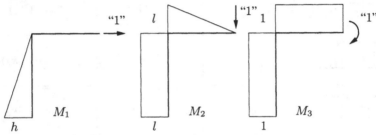

Fig. 6.5 Unit loads and bending moments

Knowing the bending moments, we can now obtain the deflections by using the formulas for the product integrals listed in Table 6.2

$$EJ\,\delta_h = \int M_0 M_1\,\mathrm{d}x = \frac{1}{2}\,h\cdot Fl\cdot h = \frac{1}{2}\,Flh^2,$$

$$EJ\,\delta_v = \int M_0 M_2\,\mathrm{d}x = \frac{1}{3}\,l\cdot Fl\cdot l + l\cdot Fl\cdot h = Fl^2\Big(\frac{l}{3}+h\Big),$$

$$EJ\,\varphi = \int M_0 M_3\,\mathrm{d}x = \frac{1}{2}\,1\cdot Fl\cdot l + 1\cdot Fl\cdot h = Fl\Big(\frac{l}{2}+h\Big).$$

Example 6.2:

Compute the bending moment diagram for the clamped beam shown in the figure. The right end of the beam is supported by a spring with stiffness $c = \alpha EJ/l^3$. (Take $\alpha = 12$ for the numerical calculation.)

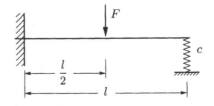

Solution:

The system is statically indeterminate of degree one. Removing the right support - or equivalently by cutting the spring - the released structure reduces to a clamped beam subjected to load F, and the redundant X_1 as external force.

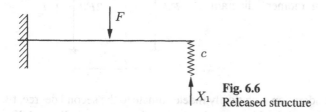

Fig. 6.6
Released structure

A spring is a typical member of a structure that exhibits extension and compression due to a given force C, and thus comparable with a bar of a truss. We therefore may either interpret these springs as specific truss members where the stiffness EA is replaced by c, or add an extra term to the relevant expressions (see Eqs. 6.39, 6.40, 6.50, 6.51), e.g.

$$W^* = \frac{1}{2} \sum_m \frac{C^2}{c_m}, \quad f_k = \sum_m \frac{C\bar{C}}{c_m}, \quad \delta_{ik} = \sum_m \frac{C_i C_k}{c_m},$$

for m different springs with stiffnesses c_m. Having this in mind, we calculate (see Table 6.2)

$$EJ\delta_{10} = -\frac{1}{6}\frac{l}{2}\left(\frac{l}{2} + 2l\right)F\frac{l}{2} = -\frac{5}{48}Fl^3,$$

$$EJ\delta_{11} = \frac{1}{3}l^3 + \frac{EJ}{c} = \frac{l^3}{3}\left(1 + \frac{3}{\alpha}\right),$$

and further from Eq. (6.53)

$$X_1 = -\frac{\delta_{10}}{\delta_{11}} = \frac{5}{16}\frac{F\alpha}{3+\alpha}.$$

We realize that the redundant X_1 takes values between:

$$X_1\big|_{\alpha=0} = 0, \quad X_1\big|_{\alpha\to\infty} = \frac{5}{16}F,$$

where $\alpha = 0$ stands for the statically determinate clamped beam, and $\alpha \to \infty$ for the statically indeterminate beam with a fixed support at the right end.

Taking the value $\alpha = 12$, we get $X_1 = \frac{1}{4}F$, and thus from Eq. (6.54) the moment diagram

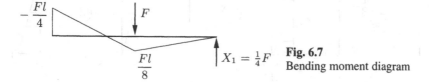

Fig. 6.7
Bending moment diagram

Example 6.3:

The plane frame shown in the figure is subject to a vertical distributed load q. The
members of the frame have constant flexural rigidities EJ, and $2EJ$, respectively.
Determine the bending moment diagrams
for this frame.

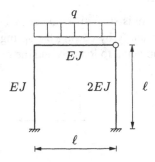

Solution:

The system under consideration is statically indeterminate to the second degree, two
redundants must be selected. In this example, the bending moments X_1 and X_2 at
the fixed supports are chosen. These bending moments can be removed from the
frame by inserting hinges at the supports, thereby producing a released structure
consisting of a pin-jointed frame.

Other choices for the two redundants are also
possible. E.g., we could have selected the
bending moment and the horizontal reaction
at the left fixed support, introducing a simple
support.

However, dealing with more complex problems (with more redundants), and having in
mind the numerical problems that may occur from the solution of Eqs. (6.53), it is always helpful to choose the released structure
as 'stiff' as possible.

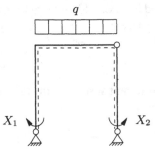

Using the diagrams for M_0, M_1, and M_2, we determine

$$EJf_{10} = \frac{1}{24} q\ell^3, \quad EJf_{20} = \frac{1}{24} q\ell^3,$$

$$EJ\delta_{11} = \frac{4}{3} \ell, \quad EJ\delta_{12} = EJ\delta_{21} = \frac{5}{6} \ell, \quad EJ\delta_{22} = \frac{5}{6} \ell,$$

where we have taken into account the fact that the right member has the (double)
flexural rigidity $2EJ$.

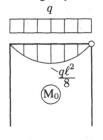

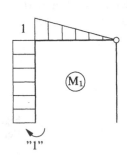

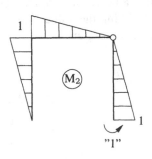

With these influence coefficients, from Eq. (6.53) we get two equations for the two unknowns X_1 and X_2

$$\frac{4}{3}\ell X_1 + \frac{5}{6}\ell X_2 = -\frac{1}{24}q\ell^3,$$

$$\frac{5}{6}\ell X_1 + \frac{5}{6}\ell X_2 = -\frac{1}{24}q\ell^3,$$

with the solutions

$$X_1 = 0, \quad X_2 = -\frac{1}{20}q\ell^2.$$

The final diagram for the bending moment is determined according to Eq. (6.54).

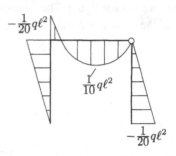

6.7 Exercises to Chapter 6

Problem 6.1:

Compute the strain energy function W of the truss given in the figure. All bars of the truss have the same cross-sectional area A and Young's modulus E. Compute the vertical, and horizontal displacements of point P. Assume linear elastic material behaviour.

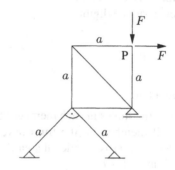

Problem 6.2:

Compute the vertical displacement of point P. Assume linear elastic material behaviour. All bars of the truss have the same cross-sectional area A and Young's modulus E.

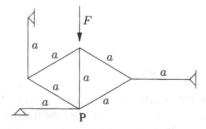

Problem 6.3:

Compute the vertical, and horizontal displacements of point P. The flexural rigidity of the beam is EJ and is constant.

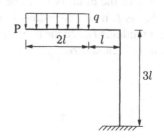

Problem 6.4:

Compute the moment diagram of the system given in the figure. The flexural rigidity of the system is EJ.

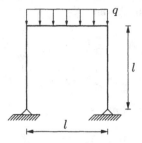

Problem 6.5:

Compute the moment diagram of the system. The flexural rigidity of the system is EJ. Assume that all deformations due to normal forces are negligible.

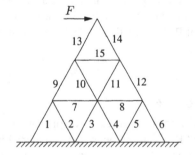

Problem 6.6:

Compute the forces in each member of the truss. All members of the truss have the same length a, cross-sectional area A, and Young's modulus E.

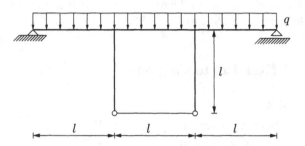

Problem 6.7:

For the system shown in the figure, compute the force in the bar after the temperature of the bar increases by θ_0. The flexural rigidity of the beam is EJ. The length of the bar is l, its cross-sectional area A, Young's modulus E, and the linear heat expansion coefficient is α.

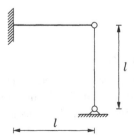

Problem 6.8:

For the system shown in the figure, compute

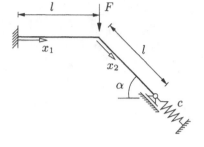

1. the reactions
2. normal-force, shear-force and bending-moment diagrams.

The flexural rigidity of the beam is EJ.

$\alpha = 45^0$, $EJ/c = l^3/3$

Problem 6.9:

For the system shown in the figure, compute the relative angle of rotation on both sides of the hinge P. Take the flexural rigidity of the beam to be EJ.

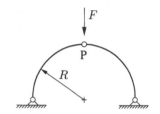

Problem 6.10:

For the system given in the figure, compute

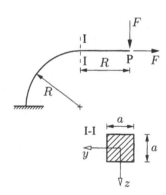

1. the angle of rotation at point P and
2. the principal stresses and planes for the centroid of the cross section I-I.

Take the flexural rigidity of the beam to be EJ. Assume that all deformations due to normal forces are negligible.

Problem 6.11:

Compute the forces in each member of the given truss. All members of the truss, save member 7, have the same cross-sectional area A, and Young's modulus E.

$$(AE)_7 = (2 \cdot AE)/ \left[(2 + 3\sqrt{2})\right]$$

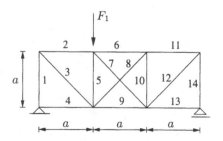

7. Two-dimensional Problems

7.1 Plane Stress and Plane Strain

Consider a thin plate of uniform thickness, made of an elastic material (see Fig. 7.1). Assume that all forces applied to its boundary or in its interior are distributed uniformly across the thickness h.

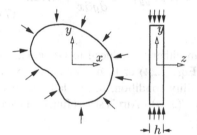

Fig. 7.1
Thin plate

Evidently, the stresses σ_{zz}, σ_{zx}, σ_{zy} vanish on both faces of the plate, and it is unlikely that they assume substantial values in the interior. Therefore their presence may be neglected entirely. Additionally, it is plausible to assume that the other three components σ_{xx}, σ_{yy}, σ_{yx} do not depend on z. The stress system described by these assumptions is called a *plane stress*.

Consider, second, a very long cylinder or prism with generators parallel to the z axis. Assume that all the surface loads and mass forces acting on this body are perpendicular to the z axis and independent of z. Furthermore, assume that the ends of the cylinder are prevented from moving in the z direction, but are otherwise free. Then the strains ε_{zz}, ε_{zx}, ε_{zy} vanish, and the other three are independent of z. Such a body is said to be in a state of *plane strain*.

Both descriptions, however, give approximate theories. Whereas in the plane stress state the stresses in the z direction are neglected, in the plane strain state self-equilibrating stresses σ_{zz} at both ends of the body are needed, which are also often neglected. Nevertheless, both descriptions yield good results in their different cases.

If a linear elastic material is presumed, Hooke's law in two dimensions is given in Section 2.52.5. We also mentioned the connection between both cases with Eqs. (2.38) and (2.39). Moreover, in both cases the equilibrium conditions (1.63) reduce to

$$\frac{\partial \sigma_{xx}}{\partial x} + \frac{\partial \sigma_{yx}}{\partial y} + \rho b_x = 0,$$

$$\frac{\partial \sigma_{xy}}{\partial x} + \frac{\partial \sigma_{yy}}{\partial y} + \rho b_y = 0,$$

(7.1)

and from the six compatibility equations (1.53, 1.54), only one is non trivial in the two dimensional case

$$\frac{\partial^2 \varepsilon_{yy}}{\partial x^2} + \frac{\partial^2 \varepsilon_{xx}}{\partial y^2} - 2\frac{\partial^2 \varepsilon_{xy}}{\partial y \partial x} = 0.$$

(7.2)

Introducing here Hooke's law, we can express Eq. (7.2) as function of the stresses

$$\frac{\partial^2}{\partial x^2}\{\sigma_{yy} - \nu\sigma_{xx}\} + \frac{\partial^2}{\partial y^2}\{\sigma_{xx} - \nu\sigma_{yy}\} - 2(1+\nu)\frac{\partial^2 \sigma_{xy}}{\partial y \partial x} + \alpha E \Delta\Theta = 0$$

(7.3)

for the plane stress state, and

$$\frac{\partial^2}{\partial x^2}\{(1-\nu)\sigma_{yy} - \nu\sigma_{xx}\} + \frac{\partial^2}{\partial y^2}\{(1-\nu)\sigma_{xx} - \nu\sigma_{yy}\} - 2\frac{\partial^2 \sigma_{xy}}{\partial y \partial x} + \alpha E \Delta\Theta = 0$$

(7.4)

for the plane strain state.

The field equations of both problems can either be reduced to two equations for the two displacements u_x and u_y (see Eqs. 2.24) or to one equation for the stress components using the above compatibility condition. In this latter case we may introduce a scalar valued stress function $F(x,y)$ - Airy's stress function - such that with

$$\sigma_{xx} = \frac{\partial^2 F}{\partial y^2}, \quad \sigma_{yy} = \frac{\partial^2 F}{\partial x^2}, \quad \sigma_{xy} = -\frac{\partial^2 F}{\partial y \partial x}$$

(7.5)

the equilibrium equations (7.1) are identically satisfied, when we neglect the body force. Introducing now Eqs. (7.5) into Eq. (7.3), we obtain the biharmonic equation

$$\boxed{\Delta\Delta F = \frac{\partial^4 F}{\partial x^4} + 2\frac{\partial^4 F}{\partial x^2 \partial y^2} + \frac{\partial^4 F}{\partial y^4} = -\alpha E \Delta\Theta.}$$

(7.6)

We note that the reduction for plane strain gives

$$\Delta\Delta F = -\frac{\alpha E}{1-\nu}\Delta\Theta.$$

(7.7)

which using Eq. (2.38) may also be directly deduced from Eq. (7.6). Moreover, for stationary temperature fields and problems without heat sources, these temperature fields follow the heat equation, which here reduces to $\Delta\Theta = 0$, and thus we find

$$\Delta\Delta F = 0$$

(7.8)

for both cases.

7.2 Body Forces Derived From a Potential

When the body-force field can be described as the gradient of a potential Φ, that is, when

$$\rho\, b_x = \frac{\partial \Phi}{\partial x}, \quad \rho\, b_y = \frac{\partial \Phi}{\partial y}, \tag{7.9}$$

the equations of equilibrium (7.1) may be written as

$$\frac{\partial}{\partial x}(\sigma_{xx} + \Phi) + \frac{\partial \sigma_{yx}}{\partial y} = 0,$$

$$\frac{\partial \sigma_{xy}}{\partial x} + \frac{\partial}{\partial y}(\sigma_{yy} + \Phi) = 0. \tag{7.10}$$

They are satisfied when we set

$$\sigma_{xx} + \Phi = \frac{\partial^2 F}{\partial y^2}, \quad \sigma_{yy} + \Phi = \frac{\partial^2 F}{\partial x^2}, \quad \sigma_{xy} = -\frac{\partial^2 F}{\partial y \partial x}. \tag{7.11}$$

Introducing these relations into the compatibility equation (7.3) for plane stress, we find

$$\boxed{\Delta\Delta F = (1 - \nu)\Delta\Phi.} \tag{7.12}$$

In the particular case that Φ is a harmonic function, satisfying the Laplace equation $\Delta\Phi = 0$, Eq. (7.12) as well as its counterpart for plane strain

$$\Delta\Delta F = \frac{1 - 2\nu}{1 - \nu}\Delta\Phi$$

reduces to the homogeneous biharmonic equation (7.8), but with Eq. (7.11) for the stresses.

7.3 Plane Problems in Polar Coordinates

Sometimes, especially in those cases where the structures under consideration show axial symmetry, it is more convenient to introduce polar coordinates r, φ. The biharmonic equation (7.6) then becomes

$$\boxed{\begin{aligned} \Delta\Delta F = {}& \frac{\partial^4 F}{\partial r^4} + \frac{2}{r}\frac{\partial^3 F}{\partial r^3} - \frac{1}{r^2}\frac{\partial^2 F}{\partial r^2} + \frac{1}{r^3}\frac{\partial F}{\partial r} + \frac{1}{r^4}\frac{\partial^4 F}{\partial \varphi^4} + \\ &+ \frac{2}{r^2}\frac{\partial^4 F}{\partial r^2 \partial \varphi^2} - \frac{2}{r^3}\frac{\partial^3 F}{\partial r \partial \varphi^2} + \frac{4}{r^4}\frac{\partial^2 F}{\partial \varphi^2} = -\alpha E \Delta\Theta, \end{aligned}} \tag{7.13}$$

where

$$\Delta = \frac{\partial^2}{\partial r^2} + \frac{1}{r}\frac{\partial}{\partial r} + \frac{1}{r^2}\frac{\partial^2}{\partial \varphi^2}, \tag{7.14}$$

is the (two dimensional) Laplace operator for polar coordinates.

The relations (7.5) between Airy's stress function and the stresses change to

$$\sigma_{rr} = \frac{1}{r^2}\frac{\partial^2 F}{\partial \varphi^2} + \frac{1}{r}\frac{\partial F}{\partial r},$$

$$\sigma_{\varphi\varphi} = \frac{\partial^2 F}{\partial r^2}, \tag{7.15}$$

$$\sigma_{r\varphi} = -\frac{\partial}{\partial r}\left\{\frac{1}{r}\frac{\partial F}{\partial \varphi}\right\},$$

and the kinematical relations are introduced as in Section 1.31.3 (Eqs. 1.49, 1.50).

The analysis is particularly simple when the stress distribution is symmetrical about the origin of the coordinates. Then all stresses, strains, displacements, and the stress function F depend on r alone, and all partial derivatives with respect to φ are zero.

When there are no body forces, Eq. (7.13) reduces to the nonhomogeneous ordinary differential (Eulers's) equation

$$\boxed{r^4 F''''(r) + 2r^3 F'''(r) - r^2 F''(r) + r F'(r) = -\alpha E r^2 (r^2 \Theta'' + r\Theta').} \tag{7.16}$$

It has the general solution

$$F(r) = c_0 + c_1 \ln r + c_2 r^2 + c_3 r^2 \ln r + F_p, \tag{7.17}$$

where the particular solution F_p is determined as a special solution of the nonhomogeneous differential equation for the given temperature field. Is this field stationary, with no heat sources, we have

$$r^2 \Theta'' + r\Theta' = 0, \tag{7.18}$$

and finally again $F_p = 0$.

The coefficients c_i in Eq. (7.17) are arbitrary constants. The corresponding stresses follow from Eq. (7.15) (for a vanishing particular solution)

$$\sigma_{rr} = \frac{1}{r}\frac{\partial F}{\partial r} = \frac{c_1}{r^2} + 2c_2 + c_3(1 + 2\ln r),$$

$$\sigma_{\varphi\varphi} = \frac{\partial^2 F}{\partial r^2} = -\frac{c_1}{r^2} + 2c_2 + c_3(1 + 2\ln r). \tag{7.19}$$

The displacements are obtained by integrating Eqs. (1.49, 1.50). From a discussion of this general solution, we may deduce that the c_3 term in Eq. (7.17) has to vanish whenever the structure has a closed form (in circumferential direction), and that $c_1 = 0$, if the plate (or the cylinder) does not have a hole at $r = 0$.

Example 7.1:

A long thick-walled cylinder with dimensions $r = a$ (inner radius), and $r = b$ (outer radius), is subject to internal pressure p from a hot gas. Due to heating a stationary temperature field of linear distribution $\Theta = \vartheta_0 + \vartheta_1 r$ may be assumed. Determine the stresses in the cylinder.

$a = 10$ cm, $b = 20$ cm, $p = 100$ MPa, $E = 2.1 \cdot 10^5$ MPa, $\nu = 0.3$,
$\alpha = 1.2 \cdot 10^{-6}$ K^{-1}, $\Theta(a) = 400$ K, $\Theta(b) = 100$ K.

Solution:

The nonhomogeneous differential equation for plane strain (7.7) takes the form

$$\Delta\Delta F = -\frac{\alpha E}{1-\nu}\frac{\vartheta_1}{r}.$$

Having determined the particular solution

$$F_p = -\frac{\alpha E}{1-\nu}\frac{\vartheta_1}{9}r^3,$$

the general solution is ($c_3 = 0$)

$$F = c_0 + c_1 \ln r + c_2 r^2 - \frac{\alpha E}{1-\nu}\frac{\vartheta_1}{9}r^3,$$

and the stresses are (Eq. 7.19)

$$\sigma_{rr} = \frac{c_1}{r^2} + 2c_2 - \frac{\alpha E}{1-\nu}\frac{\vartheta_1}{3}r,$$

$$\sigma_{\varphi\varphi} = -\frac{c_1}{r^2} + 2c_2 - \frac{\alpha E}{1-\nu}\frac{2\vartheta_1}{3}r.$$

From the boundary conditions

$$\sigma_{rr}(a) = -p, \quad \sigma_{rr}(b) = 0,$$

we finally get

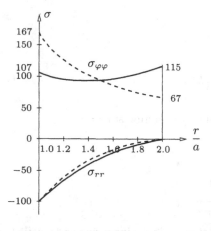

Fig. 7.2
Stresses $\sigma_{\varphi\varphi}$ and σ_{rr} vs the radius of a
thick-walled cylinder

$$\sigma_{rr} = \frac{pa^2}{b^2 - a^2}\left(1 - \frac{b^2}{r^2}\right) + \frac{\alpha E}{1-\nu}\frac{\vartheta_1}{3}\left\{\frac{a^2}{b+a}\left(1 - \frac{b^2}{r^2}\right) + b - r\right\}$$

$$\sigma_{\varphi\varphi} = \frac{pa^2}{b^2 - a^2}\left(1 + \frac{b^2}{r^2}\right) + \frac{\alpha E}{1-\nu}\frac{\vartheta_1}{3}\left\{\frac{a^2}{b+a}\left(1 + \frac{b^2}{r^2}\right) + b - 2r\right\}.$$

The results for the given quantities, and the different influences of internal pressure and heating on the stresses is shown with Fig. 7.2. The dashed lines herein show the distribution of the radial and circumferential stresses due to internal pressure. The remarkable influence of the heating on this distribution can be seen from the full lines giving the sum of both contributions.

Example 7.2:

In order to reduce the peak value of the hoop stress $\sigma_{\varphi\varphi}$, a ring may be built up of two concentric rings. In the unstressed condition, the outer diameter of the inner ring is larger than the inner diameter of the outer ring, the difference being 2δ. The parts are assembled after heating the outer ring. Determine the pressure p_c between the two faces at $r = c$ after cooling.

Solution:

From the kinematical relations and Hooke's law, we find (from Eq. 7.19, with $c_3 = 0$)

$$u_r = \frac{r}{E}\{\sigma_{\varphi\varphi} - \nu\sigma_{rr}\} = \frac{r}{E}\left\{2(1-\nu)c_2 - (1+\nu)\frac{c_1}{r^2}\right\}.$$

The shrink-fit pressure p_c can be calculated by applying first the boundary conditions for the outer ring (e) and the inner ring (i)

(e): $\sigma_{rr}(c) = -p_c,$ $\sigma_{rr}(b) = 0,$
(i): $\sigma_{rr}(c) = -p_c,$ $\sigma_{rr}(a) = 0,$

to the equations for the stresses. With this information, we describe the deformations in radial direction

$$u_{r(e)} = \frac{r}{E}\frac{p_c c^2}{b^2 - c^2}\left\{1 - \nu + (1+\nu)\frac{b^2}{r^2}\right\},$$

and

$$u_{r(i)} = \frac{r}{E}\frac{p_c c^2}{a^2 - c^2}\left\{1 - \nu + (1+\nu)\frac{b^2}{r^2}\right\}.$$

Both solutions have to satisfy the compatibility condition at $r = c$

$$u_{r(e)} - u_{r(i)} = \delta.$$

From this condition, we calculate

$$p_c = \frac{E\delta}{c}\frac{(c^2 - a^2)(b^2 - c^2)}{(b^2 - a^2)[(1-\nu)c^2 + (1+\nu)b^2]}.$$

This pressure produces compressive hoop stresses in the inner ring and tensile hoop stresses in the outer ring. It is obvious that for internal pressure the maximum hoop stress in the built-up ring is smaller than in a ring without prestress.

Example 7.3:

The outer edge of a circular disc with radius a and uniform thickness is heated by a temperature θ_0, and then kept at this temperature. The stationary temperature field in the disc may be described – with sufficient accuracy – by a quadratic function

$$\Theta = \theta_0 - m(a^2 - r^2).$$

Compute the distribution of the stresses in the disc, and the displacements of the edge, with

$\theta_0 = 200$ K, $a = 40$ cm, $m = 3/40$ K/cm^2,
$\alpha = 1.2 \cdot 10^{-5}$ K^{-1}, $E = 2 \cdot 10^5$ MPa.

Solution:

The nonhomogeneous differential equation for plane stress takes the form (Eq. 7.13)

$$\Delta\Delta F = -\alpha E\Delta\Theta = -4m\alpha E,$$

with the particular solution

$$F_p = -\frac{1}{16} m\alpha Er^4.$$

Having in mind the remarks which have been made in the paragraph following Eq. (7.19), the general solution is ($c_1 = c_3 = 0$)

$$F = c_0 + c_2 r^2 - \frac{1}{16} m\alpha Er^4,$$

and thus the stresses (Eq. 7.19)

$$\sigma_{rr} = 2c_2 - \frac{1}{4}\alpha Emr^2,$$

$$\sigma_{\varphi\varphi} = 2c_2 - \frac{3}{4}\alpha Emr^2.$$

The remaining constant c_2 is determined from the boundary condition

$$\sigma_{rr}(a) = 0, \quad \rightarrow \quad c_2 = \frac{1}{8}\alpha Ema^2,$$

such that the stresses become

$$\sigma_{rr} = \frac{1}{4}\alpha Em(a^2 - r^2),$$

$$\sigma_{\varphi\varphi} = \frac{1}{16}\alpha Em(a^2 - 3r^2).$$

The distribution of the stresses for the given numbers is shown in Fig. 7.3. It turns out that the heating produces tensile stresses of $\sigma_{rr}(0) = \sigma_{\varphi\varphi}(0) = 72$ MPa in the interior of the disc. At the edge, however, with $\sigma_{rr}(a) = 0$ and $\sigma_{\varphi\varphi}(a) = -144$ MPa considerable compressive stresses in circumferential direction occur, which may cause buckling of the disc.

The displacement u_r is determined from the strain in circumferential direction (Eqs. 1.49, 2.34)

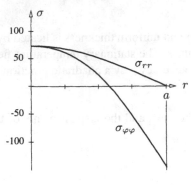

Fig. 7.3
Distribution of stresses σ_{rr} and $\sigma_{\varphi\varphi}$ vs the radius of a heated disc

$$u_r = r\varepsilon_{\varphi\varphi} = \frac{r}{E}\left(\sigma_{\varphi\varphi} - \nu\sigma_{rr}\right) + \alpha\Theta r\,,$$

and thus

$$u_r = \alpha a\left(\theta_0 - \frac{1}{2}ma^2\right) = 0.67 \text{ mm}\,.$$

7.4 Exercises to Chapter 7

Problem 7.1:

For each of the thin, rectangular disks shown in the figure, compute Airy's stress function and the stresses inside the disk. Take the thickness of each disk to be h and assume that the disks are homogeneous and isotropic.

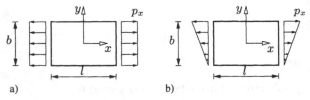

Problem 7.2:

For each of the thin, circular disks subjected to uniform pressure σ_0, compute Airy's stress function and the stresses inside the disk. Take the thickness of each disk to be h and assume that the disks are homogeneous and isotropic.

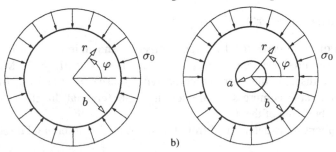

Problem 7.3:

A section of a beam of circular form is loaded by pur bending. The beam has thickness t, and its inner and outer radius are a and b, respectively. Determine Airy's stress function $F(\varphi)$. From this function, compute the stresses σ_{rr}, $\sigma_{\varphi r}$, and $\sigma_{\varphi\varphi}$. Assume plane stress conditions.

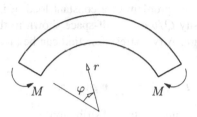

Problem 7.4:

A thin strip of a plate (middle surface $x = 0$) is subjected to a pure bending moment M without any shear force in the xz-plane as shown in the diagram. Compute the displacements $u_x(x, z)$, $u_y(x, z)$, and $u_z(x, z)$.

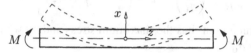

Problem 7.5:

For the simple shear problem shown in the figure, compute the displacement fields $u_x(x, y)$, and $u_y(x, y)$.

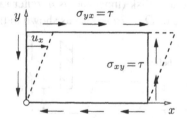

Problem 7.6:

For the clamped rectangular plate strip subjected to a uniform load of intensity Q/b at the free end, Airy's stress function is

$$F(x, y) = -\frac{3Q}{2\,bh}\left(xy - \frac{4\,xy^3}{3\,h^2}\right).$$

Compute from this function the distribution of the stresses σ_{xx} and σ_{xy} in the strip.

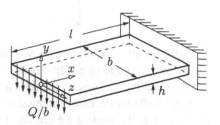

Problem 7.7:

For the problem of a constant load of intensity Q/b on a half-space shown in the figure, Airy's stress function can be taken to be

$$F(r, \varphi) = \frac{Q}{\pi b} r \varphi \cos \varphi.$$

Compute the stress distribution.

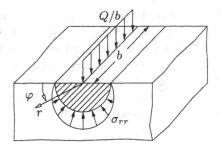

Problem 7.8:

For a plane stress strip with a circular hole (radius R) subjected to a constant stress σ_0 far away from the hole, Airy's stress function is

$$F(r, \varphi) = \frac{\sigma_0 R^2}{4} \left[\left(\frac{r}{R} \right)^2 - 2 \ln \frac{r}{R} - \left(\frac{r}{R} \right)^2 \left(1 - \left(\frac{R}{r} \right)^2 \right)^2 \cos 2\varphi \right].$$

Compute the stress distribution. Which point has the maximum stress and by what magnitude is this stress greater than σ_0?

Problem 7.9:

A circular disk (inner radius a, outer radius b) is to be shrunk around a steel shaft (diameter $D = 2a + \Delta d$) as shown in the diagrams.

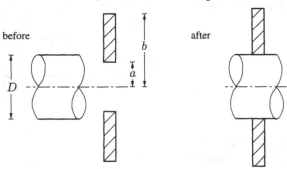

1. Calculate the temperature above room temperature to which the disk must be raised in order to be easily shrunk on the shaft.
2. Compute the stresses in the disk at room temperature (i.e. after the disk has cooled off) for the case that the shaft is rigid.
3. How does the stress state change if one assumes a plane and axial symmetric stress state in the shaft?
4. Solve the problem for a plane strain state.

   $a = 10$ cm, $E_D = 7 \cdot 10^4$ MPa, $\nu_D = 0.3$, $\Delta d = 0.5$ mm,
   $b = 30$ cm, $E_S = 2 \cdot 10^5$ MPa, $\nu_S = 0.3$, $\alpha = 2.4 \cdot 10^{-5}$ K^{-1}.

8. Plates and Shells

8.1 General Remarks

In this chapter, we will consider 2D structures, and assume that the shape of the structure is described by its middle surface and thickness h perpendicular to this middle surface. We assume further that the thickness is small compared to the dimensions of the middle surface, which means that here we restrict our considerations to thin-walled 2D structures. According to the shape of the middle surface, and the loading of the structure, we distinguish between:

(i) Plates with plane middle surface, and forces acting parallel to this middle surface. In the sequel we will call this a disk.

(ii) Plates with plane middle surfaces, and forces acting perpendicular to the middle surfaces, and

(iii) Shells with curved middle surfaces.

Under certain assumptions, these problems can be described with the two coordinates of the middle surface.

Here, just as we did for beams, where we introduced stress resultants to reduce (map) the three-dimensional stress state to the one-dimensional beam axis, we define stress resultants by integrating over the entire thickness h. With cartesian coordinates x, y, z introduced such that the plane $z = 0$ coincides with the middle surface, we may define the following stress resultants (per unit length)

$$n_{xx} = \int_{-\frac{h}{2}}^{+\frac{h}{2}} \sigma_{xx}\, dz\,, \qquad n_{xy} = \int_{-\frac{h}{2}}^{+\frac{h}{2}} \sigma_{xy}\, dz\,, \qquad n_{yy} = \int_{-\frac{h}{2}}^{+\frac{h}{2}} \sigma_{yy}\, dz\,,$$

$$q_{x} = \int_{-\frac{h}{2}}^{+\frac{h}{2}} \sigma_{xz}\, dz\,, \qquad q_{y} = \int_{-\frac{h}{2}}^{+\frac{h}{2}} \sigma_{yz}\, dz\,, \tag{8.1}$$

$$m_{xx} = \int_{-\frac{h}{2}}^{+\frac{h}{2}} z\sigma_{xx}\, dz\,, \qquad m_{xy} = \int_{-\frac{h}{2}}^{+\frac{h}{2}} z\sigma_{xy}\, dz\,, \qquad m_{yy} = \int_{-\frac{h}{2}}^{+\frac{h}{2}} z\sigma_{yy}\, dz\,.$$

We also may define zero-th and first moments of a temperature change Θ, which are calculated to give

$$n^\theta = \int_{-\frac{h}{2}}^{+\frac{h}{2}} E\alpha\Theta \, dz = E\alpha\theta_0 \, h \, ,$$

$$m^\theta = \int_{-\frac{h}{2}}^{+\frac{h}{2}} z E\alpha\Theta \, dz = E\alpha\theta_1 \frac{h^3}{12} \, .$$

(8.2)

for a temperature field

$$\Theta(x, y, z) = \theta_0(x, y) + \theta_1(x, y)z \, ,$$

(8.3)

linearly varying with z. The normal stress in z direction σ_{zz} is neglected, as we usually do for thin-walled structures.

8.2 Disks

In the preceding chapter, we introduced a plane stress state to describe the behaviour of a thin plate under forces applied to its boundary, and uniformly distributed across the thickness.

This is just the situation we observe with disks, if the following restrictions are assumed:

(i)   A constant thickness h,
(ii) conservative body forces, i.e. forces derivable from a potential, and
(iii) a loading uniformly distributed across the thickness.

Under these assumptions, the mean values of the stresses σ_{zx} and σ_{zy} may be neglected, and thus we find

$$q_x = q_y = 0 \, ,$$

(8.4)

and

$$m_{xx} = m_{xy} = m_{yy} = m^\theta = 0 \, ,$$

(8.5)

since due to the symmetry of the loading, the components σ_{xx}, σ_{xy}, σ_{yy} as well as the temperature change Θ do not depend on z. This means that

$$n_{xx} = h\sigma_{xx} \, , \quad n_{xy} = n_{yx} = h\sigma_{xy} \, , \quad n_{yy} = h\sigma_{yy} \, ,$$

(8.6)

and thus the behaviour of the disk is described by the equations of plane stress, e.g. Eq. (7.6).

Example 8.1:
Determine the stresses in a solid rotating circular disk of radius a and uniform thickness.

Solution:

For a disk of uniform thickness and of specific mass ρ, rotating at constant angular velocity ω, d'Alemberts inertia force per unit volume is

$$F = \rho \omega^2 r ,$$

directed outward. It may be derived from a potential

$$\Phi = \frac{1}{2} \rho \omega^2 r^2 .$$

The right-hand side of Eq. (7.12) is, in this case,

$$(1 - \nu) \Delta \Phi = (1 - \nu) \left\{ \frac{\partial^2 \Phi}{\partial r^2} + \frac{1}{r} \frac{\partial \Phi}{\partial r} \right\} = 2(1 - \nu) \rho \omega^2 ,$$

and, since $\Delta\Delta F$ is given by Eq. (7.13), the differential equation

$$F'''' + \frac{2}{r} F''' - \frac{1}{r^2} F'' + \frac{1}{r^3} F' = 2(1 - \nu)\rho \omega^2$$

must be solved.

Obviously,

$$F_p = \frac{1 - \nu}{32} \rho \omega^2 r^4$$

is a particular solution. Adding the complementary solution, we have the general solution (Eq. 7.12)

$$F(r) = c_0 + c_1 \ln r + c_2 r^2 + c_3 r^2 \ln r + \frac{1 - \nu}{32} \rho \omega^2 r^4 .$$

Uniqueness of displacements requires that $c_3 = 0$. What is left yields the stress field

$$\sigma_{rr} = \frac{c_1}{r^2} + 2c_2 - \frac{3 + \nu}{8} \rho \omega^2 r^2 ,$$

$$\sigma_{\varphi\varphi} = -\frac{c_1}{r^2} + 2c_2 - \frac{1 + 3\nu}{8} \rho \omega^2 r^2 .$$

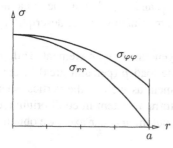

Fig. 8.1
Stresses in a rotating disc

For a solid disk, regularity of the stress field at $r = 0$ requires that $c_1 = 0$. c_2 is determined from the boundary condition $\sigma_{rr} = 0$ at $r = a$. It yields

$$c_2 = \frac{3+\nu}{16}\rho\omega^2 a^2,$$

and the stresses are

$$\sigma_{rr} = \frac{3+\nu}{8}\rho\omega^2(a^2 - r^2),$$

$$\sigma_{\varphi\varphi} = \frac{3+\nu}{8}\rho\omega^2 a^2 - \frac{1+3\nu}{8}\rho\omega^2 r^2.$$

The distribution of these stresses is shown with Fig. 8.1. Both stresses reach their maxima at the center of the disk, where

$$\sigma_{rr} = \sigma_{\varphi\varphi} = \frac{3+\nu}{8}\rho\omega^2 a^2.$$

8.3 Thin Plates

8.3.1 Fundamental equations

In the theory of thin plates, it is customary to make the following assumptions:

(1) The plate is initially flat.
(2) The material is elastic, homogeneous, and isotropic.
(3) Thickness is small compared to area dimensions.
(4) Slope of the deflection surface is small compared to unity.
(5) Deformation is such that straight lines initially normal to the middle surface remain straight and normal to that surface (Bernoulli's hypothesis).
(6) Strains in the middle surface, arising from the deflection, are neglected compared to strains due to bending.
(7) Deflection of the plate occurs by virtue of displacements of points in the mid-surface normal to its initial plane.

Due to these assumptions, and the fact that only forces acting perpendicular to the middle surface are considered, we find

$$n_{xx} = n_{xy} = n_{yy} = n^\theta = 0.\qquad(8.7)$$

We thus realize that the equations for disks and plates are decoupled. The equations for disks contain the stress resultants n_{xx}, n_{xy}, n_{yy}, and n^θ, whereas the remaining stress resultants q_x, q_y, m_{xx}, m_{xy}, m_{yy}, and m^θ are related with the description of plate bending.

Let us consider a small element of the plate generated on an element $dx\,dy$ of the middle surface. This element is subject to the load p (per unit area) in the z direction. The moments acting on this small element as well as the vertical shear stress resultants q_x and q_y, and the load p, will form a system in equilibrium (see Fig. 8.2). Considering equilibrium of moments about the y and x axes, we obtain

$$\frac{\partial m_{xx}}{\partial x} + \frac{\partial m_{yx}}{\partial y} - q_x = 0,$$

$$\frac{\partial m_{xy}}{\partial x} + \frac{\partial m_{yy}}{\partial y} - q_y = 0.\qquad(8.8)$$

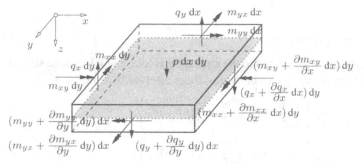

Fig. 8.2 Stress resultants acting on a plate element

Consideration of vertical equilibrium gives

$$\frac{\partial q_x}{\partial x} + \frac{\partial q_y}{\partial y} + p(x,y) = 0.$$

(8.9)

Substitution of the vertical shear forces from Eqs. (8.8) into Eq. (8.9) leads to the following fundamental equation

$$\boxed{\frac{\partial^2 m_{xx}}{\partial x^2} + 2\frac{\partial^2 m_{xy}}{\partial x \partial y} + \frac{\partial^2 m_{yy}}{\partial y^2} = -p(x,y).}$$

(8.10)

Equation (8.10) is called the moment equilibrium equation of plates. We note that this equation is independent of any material behaviour.

8.3.2 Kirchhoff's theory

To solve this equation, we now introduce displacements, strains, and the material description of a linearly elastic body (Hooke's law).

The middle plane of the plate, before deflection, is assumed to lie at $z = 0$. When the plate is being deformed, points in the middle surface move parallel to the z axis, and, at the point x, y, the deflection is $u_z = w$.

The bending moments will depend on the stresses in the plate, and these, in turn, depend on the strains. Since the horizontal displacements u_x and u_y in the x and y directions depend on the z coordinate and the slope components, we have

$$u_x = -z\frac{\partial w}{\partial x}, \quad u_y = -z\frac{\partial w}{\partial y},$$

(8.11)

and the strains are

$$\varepsilon_{xx} = \frac{\partial u_x}{\partial x} = -z\frac{\partial^2 w}{\partial x^2},$$

$$\varepsilon_{yy} = \frac{\partial u_y}{\partial y} = -z\frac{\partial^2 w}{\partial y^2},$$

(8.12)

$$\varepsilon_{xy} = \frac{1}{2}\left\{\frac{\partial u_y}{\partial x} + \frac{\partial u_x}{\partial y}\right\} = -z\frac{\partial^2 w}{\partial x \partial y},$$

as well as

$$\varepsilon_{zz} = \varepsilon_{xz} = \varepsilon_{yz} = 0. \tag{8.13}$$

Stresses σ_{xx}, σ_{xy}, σ_{yy} are related to the strains by Hooke's law for plane stress (see Eqs. 2.34). In terms of the plate curvatures, these stresses become

$$\sigma_{xx} = -\frac{Ez}{1-\nu^2}\left\{\frac{\partial^2 w}{\partial x^2} + \nu\frac{\partial^2 w}{\partial y^2} + (1+\nu)\theta_1\right\},$$

$$\sigma_{yy} = -\frac{Ez}{1-\nu^2}\left\{\frac{\partial^2 w}{\partial y^2} + \nu\frac{\partial^2 w}{\partial x^2} + (1+\nu)\theta_1\right\}, \tag{8.14}$$

$$\sigma_{xy} = -\frac{Ez}{1+\nu}\frac{\partial^2 w}{\partial x\partial y}.$$

We note that herein $\theta_1 = \theta_1(x,y)$ is the gradient of the temperature field Θ (see Eq. 8.3) in z direction.

By integrating the stress components over the thickness h of the plate, we obtain the bending and twisting moments per unit length (see Eqs. 8.1 3)

$$m_{xx} = -B\left\{\frac{\partial^2 w}{\partial x^2} + \nu\frac{\partial^2 w}{\partial y^2} + (1+\nu)\theta_1\right\},$$

$$m_{yy} = -B\left\{\frac{\partial^2 w}{\partial y^2} + \nu\frac{\partial^2 w}{\partial x^2} + (1+\nu)\theta_1\right\}, \tag{8.15}$$

$$m_{xy} = -B(1-\nu)\frac{\partial^2 w}{\partial x\partial y},$$

where

$$B = \frac{Eh^3}{12(1-\nu^2)} \tag{8.16}$$

is the flexural rigidity or bending stiffness of the plate. Introducing these equations into the moment equilibrium equation leads to the basic differential equation of plate theory (Kirchhoff's theory)

$$\Delta\Delta w = \frac{\partial^4 w}{\partial x^4} + 2\frac{\partial^4 w}{\partial x^2\partial y^2} + \frac{\partial^4 w}{\partial y^4} = \frac{p(x,y)}{B} - (1+\nu)\Delta\theta_1. \tag{8.17}$$

Again, this is a nonhomogeneous biharmonic equation.

Having determined the bending and twisting moments, the shear force resultants can be calculated from Eq. (8.8) to give

$$q_x = -B\frac{\partial}{\partial x}\left\{\frac{\partial^2 w}{\partial x^2} + \frac{\partial^2 w}{\partial y^2} + (1+\nu)\theta_1\right\},$$

$$q_y = -B\frac{\partial}{\partial y}\left\{\frac{\partial^2 w}{\partial x^2} + \frac{\partial^2 w}{\partial y^2} + (1+\nu)\theta_1\right\}. \tag{8.18}$$

8.3.3 Boundary conditions

The differential equation (8.17) is of 4th order, and thus the number of conditions that can be satisfied at every boundary is two. They may involve either the deflection and slope or the shear forces and moments, or a combination of these quantities. If, however, dynamical conditions are prescribed, e.g. at a free edge, conditions for three resultants have to be fulfilled, namely q_n, m_{nn}, m_{ns}, where here the subscripts n and s are used to indicate normal and tangential directions of the boundary.

Following a suggestion of Thomson and Tait, this problem can be overcome. Just as a distribution of externally applied moments on a beam can be regarded as a (statically equivalent) distribution of normal loads, so can the distribution of twisting moments near an edge of the plate be regarded as producing an equivalent vertical force per unit length. The total vertical force (per unit length) q_n^* on a boundary thus consists of the sum of this force and the vertical shear force q_n:

$$q_n^* = q_n + \frac{\partial m_{ns}}{\partial s} = -B \frac{\partial}{\partial n} \left\{ \frac{\partial^2 w}{\partial n^2} + (2 - \nu) \frac{\partial^2 w}{\partial s^2} + (1 + \nu)\theta_1 \right\}. \quad (8.19)$$

Wherever the boundary of the plate has a corner, there may be a concentrated force Q^*, e.g.

$$Q^* = -2m_{xy} = -2m_{yx} \quad (8.20)$$

at a rectangular corner, where the edges coincide with the x and y directions.

The various types of boundary conditions for isothermal problems are now listed in Table 8.1.

8.3.4 Axially symmetric bending of circular plates

When Eq. (8.17), for vertical equilibrium of a plate element, is transformed into polar coordinates r and φ, it has the form (see Eq. 7.13)

$$\Delta\Delta w = \frac{p(r, \varphi)}{B} - (1 + \nu)\Delta\theta_1(r, \varphi), \quad (8.21)$$

where now the Laplace operator takes the form of Eq. (7.14). For axially symmetric bending this form can be further simplified to give

$$\Delta = \frac{\partial^2}{\partial r^2} + \frac{1}{r} \frac{\partial}{\partial r}, \quad (8.22)$$

and

$$\boxed{\Delta\Delta w = w'''' + \frac{2}{r} w''' - \frac{1}{r^2} w'' + \frac{1}{r^3} w' = \frac{p(r)}{B} - (1 + \nu)\Delta\theta_1(r),} \quad (8.23)$$

where now the prime describes differentiation with respect to radius r.

Moments and shear forces simplify to

$$m_{rr} = -B\left\{w'' + \frac{\nu}{r}w' + (1+\nu)\theta_1\right\},$$

$$m_{\varphi\varphi} = -B\left\{\nu\, w'' + \frac{1}{r}w' + (1+\nu)\theta_1\right\},$$

(8.24)

and

$$q_r = m'_{rr} + \frac{1}{r}(m_{rr} - m_{\varphi\varphi})$$

$$= -B\left\{w'' + \frac{1}{r}w' + (1+\nu)\theta_1\right\}'.$$

(8.25)

a) Boundary conditions at straight edges

Type	Boundary conditions	Conditions expressed in w
Clamped edge	$w = 0$	$w = 0$
	$\dfrac{\partial w}{\partial n} = 0$	$\dfrac{\partial w}{\partial n} = 0$
Free edge	$m_{nn} = 0$	$\dfrac{\partial^2 w}{\partial n^2} + \nu\dfrac{\partial^2 w}{\partial s^2} = 0$
	$\left.\begin{array}{l} q_n = 0 \\ m_{ns} = 0 \end{array}\right\} q^* = 0$	$\dfrac{\partial^3 w}{\partial n^3} + (2-\nu)\dfrac{\partial^3 w}{\partial s^2 \partial n} = 0$
Simply supported edge	$w = 0$	$w = 0$
	$m_{nn} = 0$	$\dfrac{\partial^2 w}{\partial n^2} = 0$ or $\Delta w = 0$

b) Boundary conditions of axially symmetric loaded circular plates

Type	Boundary conditions	Conditions expressed in w
Clamped edge	$w = 0$	$w = 0$
	$w' = 0$	$w' = 0$
Free edge	$m_{rr} = 0$	$w'' + \dfrac{\nu}{r}w' = 0$
	$q_r = 0$	$w''' + \dfrac{1}{r}w'' - \dfrac{1}{r^2}w' = 0$
Simply supported edge	$w = 0$	$w = 0$
	$m_{rr} = 0$	$w'' + \dfrac{\nu}{r}w' = 0$

Table 8.1 Boundary conditions

Problems can often be solved by superposition of elementary cases. In general, the solutions of Eq. (8.23) will involve combination of a particular solution and the general solution of the homogeneous equation. The latter is known to be (see Eq. 7.17)

$$w(r) = c_0 + c_1 \ln r + c_2\, r^2 + c_3\, r^2 \ln r. \tag{8.26}$$

Since the twisting moment $m_{r\varphi}$ vanishes for axially symmetric problems, the boundary conditions can be formulated straightforwardly as also given in Table 8.1, where again these conditions only for the sake of simplicity have been restricted to isothermal situations.

Example 8.2:

Determine the maximum stresses in a clamped circular plate (radius R, thickness h) with uniform load p_0.

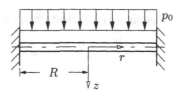

Solution:

The particular integral may be taken as

$$w_p = \frac{p_0 r^4}{64B},$$

and only terms involving c_0 and c_2 from (8.26) apply, due to the boundedness of the deflection and the bending moments at $r = 0$,

$$w = c_0 + c_2\, r^2 + \frac{p_0 r^4}{64B}.$$

Use of clamped-edge boundary conditions $w = w' = 0$ at $r = R$ leads to

$$c_0 = \frac{p_0 R^4}{64B}, \quad c_2 = -\frac{p_0 R^2}{32B},$$

and thus

$$w = \frac{p_0 R^4}{64B}\left\{1 - \frac{r^2}{R^2}\right\}^2,$$

with moments

$$m_{rr} = \frac{p_0 R^2}{16}\left\{1 + \nu - (3 + \nu)\frac{r^2}{R^2}\right\},$$

$$m_{\varphi\varphi} = \frac{p_0 R^2}{16}\left\{1 + \nu - (1 + 3\nu)\frac{r^2}{R^2}\right\},$$

and a shear force q_r of magnitude (see Eq. 8.25)

$$q_r = -\frac{1}{2}\, p_0 r.$$

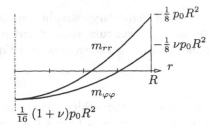

$-\frac{1}{8} p_0 R^2$

m_{rr}

$-\frac{1}{8} \nu p_0 R^2$

r

R

$m_{\varphi\varphi}$

$\frac{1}{16}(1+\nu) p_0 R^2$

Fig. 8.3
Distribution of the moments in a clamped
circular plate

The distribution of the moments is plotted in Fig. 8.3. Maximum values will be
obtained at $r = R$ with

$$m_{rr} = -\frac{p_0 R^2}{8}, \quad m_{\varphi\varphi} = -\nu \frac{p_0 R^2}{8},$$

with a maximum equivalent stress (v. Mises criterion)

$$\sigma_{eq} = \frac{3}{4} \frac{R^2}{h^2} p_0 \sqrt{1 - \nu + \nu^2}.$$

Example 8.3:
The clamped circular plate of Example 8.2 is subjected to a singular vertical force F
at the center of the plate. Compute the deflection, the shear force, and the moments.

Solution:

Since there is no distributed force acting on the plate the solution can be taken from
Eq. (8.26) alone

$$w(r) = c_0 + c_2 r^2 + c_3 r^2 \ln r,$$

where again due to the boundedness of the deflection at $r = 0$ the term involving
c_1 has been cancelled. The last term, however, has to be considered since here the
bending moments need not be bounded at $r = 0$, and for the deflection we have

$$\lim_{r \to 0} r^2 \ln r = 0.$$

The shear force q_r at $r = $ const. is (see Eq. 8.25)

$$q_r = -B \left\{ w'' + \frac{1}{r} w' \right\}'.$$

From the equilibrium condition in z-direction for a circular element with radius r
cut from the plate, and including the singular force F, we find

$$q_r \cdot 2\pi r + F = 0.$$

This condition leads us to

$$c_3 = \frac{F}{8\pi B},$$

and thus the remaining coefficients may be determined from the boundary conditions $w = w' = 0$ at $r = R$

$$c_0 = \frac{FR^2}{16\pi B}, \quad c_2 = -\frac{F}{16\pi B}(1 + 2\ln R).$$

We finally get:

$$w = \frac{FR^2}{16\pi B}\left\{1 - \frac{r^2}{R^2} + 2\ln\frac{r}{R}\frac{r^2}{R^2}\right\},$$

with moments

$$m_{rr} = -\frac{F}{4\pi}\left\{1 + (1 + \nu)\ln\frac{r}{R}\right\}, \quad m_{\varphi\varphi} = -\frac{F}{4\pi}\left\{\nu + (1 + \nu)\ln\frac{r}{R}\right\},$$

and shear force

$$q_r = -\frac{F}{2\pi r}.$$

Example 8.4:

The annular plate shown in the figure is subjected to a bending moment m_0 at its outer edge. Compute the deflection w, and the moments m_{rr} and $m_{\varphi\varphi}$.

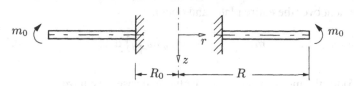

Solution:

As in the preceding example, the solution can be taken from Eq. (8.26)

$$w(r) = c_0 + c_1 \ln r + c_2 r^2 + c_3 r^2 \ln r,$$

with kinematical and dynamical boundary conditions:

$$w(R_0) = w'(R_0) = 0,$$

$$q_r(R) = 0, \quad m_{rr}(R) = m_0.$$

It turns out from Eq. (8.25) that the shear force q_r at $r = $ const. may be expressed by

$$q_r = -B\left\{w'' + \frac{1}{r}w'\right\}' = -4Bc_3\frac{1}{r}.$$

Thus, from the first dynamical condition, we immediately determine coefficient c_3

$$c_3 = -\frac{1}{4B}q_r r\bigg|_R = 0 \quad \rightarrow \quad q_r(r) \equiv 0.$$

The remaining coefficients are determined from the three remaining conditions

$$c_0 = \frac{m_0}{2B\Phi} R_0^2 (1 - 2\ln R_0), \quad c_1 = \frac{m_0}{B\Phi} R_0^2, \quad c_2 = -\frac{m_0}{2B\Phi}.$$

with

$$\Phi = 1 + \nu + (1 - \nu)\rho_0^2, \quad \rho_0 = \frac{R_0}{R}, \quad \rho = \frac{r}{R_0},$$

and the final result for the deflection

$$w = \frac{m_0}{2B\Phi} R_0^2 \left(1 - \rho^2 + 2\ln\rho\right),$$

as well as moments

$$m_{rr} = \frac{m_0}{\Phi} \left\{1 + \nu + (1 - \nu)\frac{1}{\rho^2}\right\}, \quad m_{\varphi\varphi} = \frac{m_0}{\Phi} \left\{1 + \nu - (1 - \nu)\frac{1}{\rho^2}\right\}.$$

8.3.5 Elastic energy of plates

The principle of virtual work states (see Eq. 1.85)

$$\delta W - \delta A = 0. \tag{8.27}$$

In thin-plate theory, only the elastic energy of bending and twisting is considered. From the definition of the virtual work of the internal forces, we integrate for a rectangular element over the entire plate, and obtain

$$\delta W = \int_A \left\{ -m_{xx}\delta w_{,xx} - m_{yy}\delta w_{,yy} - 2m_{xy}\delta w_{,xy} \right\} dA. \tag{8.28}$$

When Eqs. (8.15) for the moments are used (isothermal case), we have

$$\delta W = B \int_A \left\{ \Delta w(\delta w_{,xx} + \delta w_{,yy}) - \right.$$
$$\left. -(1 - \nu)(w_{,xx}\delta w_{,yy} + w_{,yy}\delta w_{,xx} - 2w_{,xy}\delta w_{,xy}) \right\} dA, \tag{8.29}$$

and thus, we determine the strain energy

$$W = \frac{1}{2} B \int_A \left\{ \Delta w^2 - 2(1 - \nu)(w_{,xx}w_{,yy} - w_{,xy}w_{,xy}) \right\} dA, \tag{8.30}$$

or, again using the expression for the moments, the complementary energy

$$W^* = \frac{1}{2B(1 - \nu^2)} \int_A \left\{ m_{xx}^2 + m_{yy}^2 + 2(1 + \nu)m_{xy}^2 - 2\nu m_{xx}m_{yy} \right\} dA. \tag{8.31}$$

The elastic energy expressions are useful in obtaining approximate solutions to plate problems.

8.4 Membrane Theory of Shells of Revolution

A shell is an object whose material is confined to the close vicinity of a curved surface, the middle surface of the shell. To locate a point on the middle surface, we use curvilinear coordinates ξ, η. In the sequel, we will discuss only those cases where the normal (membrane) forces acting in the middle surface $n_{\xi\xi}$, $n_{\eta\eta}$, $n_{\xi\eta}$ are sufficient to describe an equilibrium of the shell under given loads. This description will be called membrane theory.

Obviously, the bending and twisting moments of a thin shell must be rather small and, hence, of little influence on the over-all picture. The membrane theory is based on the assumption that they are small enough to be neglected. It meets the limits of its applicability when either external moments are applied at the edge of the shell, or the membrane deformations lead to discrepancies which can only be resolved by admitting the existence of small, but finite bending and twisting moments.

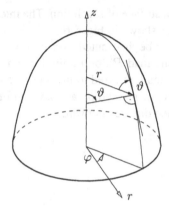

Fig. 8.4
Shell of revolution

If all moments (and thus also the shear forces) vanish, the moment equilibrium of a shell element requires that $n_{\xi\eta} = n_{\eta\xi}$.

Figure 8.4 shows a surface of revolution and the curvilinear coordinates ϑ, φ. Lines for which $\vartheta = $ const. are called parallel circles, lines $\varphi = $ const. are the meridians. We denote by r the radius of the parallel circle, by R_ϑ the radius of curvature of the meridian, and by R_φ the length of a normal between the surface and the axis of revolution.

Using ϑ and φ, we can describe the position of any point of the middle surface with these parameters, e.g. in cylindrical polar coordinates

$$r(\vartheta), \ \varphi, \ z(\vartheta).$$

For some special cases, however, e.g. for conical and cylindrical shells, the slope of the meridian ϑ is a constant and thus not suitable as a coordinate. We then use a new coordinate s along this straight generator.

Due to the symmetry of the shell, we find the following principal curvatures (see Fig. 8.5).

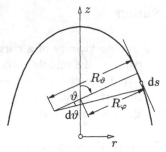

Fig. 8.5
Curvatures of shell

$$\frac{1}{R_\vartheta} = \frac{d\vartheta}{ds} = \cos\vartheta \, \frac{d\vartheta}{dr} \,, \quad \rightarrow \quad \frac{1}{R_\vartheta} = \frac{\cos\vartheta}{dr/d\vartheta} \,, \tag{8.32}$$

and

$$\frac{1}{R_\varphi} = \frac{\sin\vartheta}{r(\vartheta)} \,. \tag{8.33}$$

We now consider a shell whose middle surface is a surface of revolution. The internal forces (membrane forces) are $n_{\vartheta\vartheta}$, $n_{\varphi\varphi}$, $n_{\vartheta\varphi}$ as shown in Fig. 8.6.

External forces (loads) applied to the shell may be distributed over part of the middle surface (surface load p) or along a line (line load P), or a finite force may act at a definite point (point load **P**). The surface load, referred to the unit area of the middle surface is resolved into its components p_ϑ, p_φ, and p_n as shown in Fig. 8.6. The radial (normal) component is positive when directed outward.

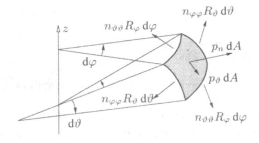

Fig. 8.6
Small element of the shell

The equilibrium of the shell element yields the equations

$$\frac{1}{R_\vartheta} \frac{\partial}{\partial\vartheta}(rn_{\vartheta\vartheta}) + \frac{\partial n_{\vartheta\varphi}}{\partial\varphi} - n_{\varphi\varphi}\cos\vartheta + rp_\vartheta = 0 \,,$$

$$\frac{1}{R_\vartheta} \frac{\partial}{\partial\vartheta}(rn_{\vartheta\varphi}) + \frac{\partial n_{\varphi\varphi}}{\partial\varphi} + n_{\vartheta\varphi}\cos\vartheta + rp_\varphi = 0 \tag{8.34}$$

as well as

$$\boxed{\frac{n_{\vartheta\vartheta}}{R_\vartheta} + \frac{n_{\varphi\varphi}}{R_\varphi} - p_n = 0 \,.} \tag{8.35}$$

They are three equations for the three unknown membrane forces. Consequently, the membrane stress problem is internally statically determinate, but it may be externally indeterminate. If the loads, including the reactions at the edge, are independent of φ, the stress system is axisymmetric, and Eqs. (8.34) reduce to

$$\frac{1}{R_\vartheta}\frac{d}{d\vartheta}(rn_{\vartheta\vartheta}) - n_{\varphi\varphi}\cos\vartheta + rp_\vartheta = 0\,,$$

$$\frac{1}{R_\vartheta}\frac{d}{d\vartheta}(rn_{\vartheta\varphi}) + n_{\vartheta\varphi}\cos\vartheta + rp_\varphi = 0\,.$$

(8.36)

Introducing here Eq. (8.35), these equations can be solved to give

$$n_{\vartheta\vartheta} = -\frac{1}{r\sin\vartheta}\int\{p_\vartheta\sin\vartheta - p_n\cos\vartheta\}\,rR_\vartheta\,d\vartheta\,,$$

$$n_{\vartheta\varphi} = -\frac{1}{r^2}\int p_\varphi\,r^2R_\vartheta\,d\vartheta\,.$$

(8.37)

Example 8.5:

A spherical cap, supported at an angle of $\vartheta_0 = 60°$, is loaded by its own weight. Determine the membrane forces, and the maximum equivalent stresses for a shell of thickness h.

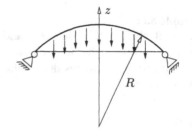

Solution:

We consider a spherical shell with $h = $ const. subjected to its own weight. For this shell, we have

$$R_\vartheta = R_\varphi = R\,,\quad r(\vartheta) = R\sin\vartheta\,.$$

The loading is given as (same for all shells of revolution)

$$p_\vartheta(\vartheta) = \rho gh\sin\vartheta$$

$$p_n(\vartheta) = -\rho gh\cos\vartheta\,.$$

With this information, we can calculate from Eqs. (8.37)

$$n_{\vartheta\vartheta}(\vartheta) = h\sigma_{\vartheta\vartheta} = -\frac{\rho ghR}{\sin^2\vartheta}\int\sin\vartheta\,d\vartheta$$

$$= -\frac{\rho ghR}{\sin^2\vartheta}(1 - \cos\vartheta) = -\frac{\rho ghR}{1 + \cos\vartheta}$$

$$n_{\varphi\varphi}(\vartheta) = h\sigma_{\varphi\varphi} = \frac{\rho ghR}{1 + \cos\vartheta} - \rho ghR\cos\vartheta = \rho ghR\left(\frac{1}{1 + \cos\vartheta} - \cos\vartheta\right).$$

The equivalent stresses are calculated according to the v. Mises criterion

$$\sigma_{eq} = \sqrt{\sigma_{\vartheta\vartheta}^2 - \sigma_{\vartheta\vartheta}\sigma_{\varphi\varphi} + \sigma_{\varphi\varphi}^2}\,.$$

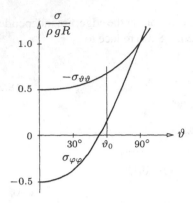

Fig. 8.7
Stresses $\sigma_{\vartheta\vartheta}$ and $\sigma_{\varphi\varphi}$ in a spherical shell

The shell is supported at $\vartheta = 60°$. Thus, the maximum equivalent stresses appear at the apex with

$$\sigma_{eq} = \rho g R \sqrt{3}/2 = 0.866\,\rho g R\,.$$

Example 8.6:
A conical shell with radius R_0 and height H is filled with water. Determine the membrane forces due to the water pressure.

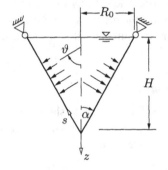

Solution:

A conical shell is considered filled with water, and subjected to the fluid pressure (γ is the specific weight of water). The weight of the shell itself is assumed to be negligible.

For a conical shell, we have $\vartheta = $ const. We therefore introduce a new coordinate s, and with the following transformations

$$R_\vartheta\,d\vartheta = ds\,, \quad (R_\vartheta \to \infty,\; d\vartheta \to 0)$$

the foregoing formulas retain their validity:

$$\vartheta = \frac{\pi}{2} - \alpha\,, \quad r(s) = s\sin\alpha\,, \quad R_\varphi(s) = \frac{r(s)}{\cos\alpha} = s\tan\alpha\,.$$

The loading is given as

$$p_n(s) = \gamma H\left(1 - \frac{s}{H}\cos\alpha\right)\,, \quad p_\vartheta(s) = 0\,.$$

We thus get from Eqs. (8.37)

$$n_{\vartheta\vartheta}(s) = h\sigma_{\vartheta\vartheta} = \frac{1}{6}\gamma H\,s\tan\alpha\left(3 - 2\frac{s}{H}\cos\alpha\right)$$

$$n_{\varphi\varphi}(s) = h\sigma_{\varphi\varphi} = \gamma H\,s\tan\alpha\left(1 - \frac{s}{H}\cos\alpha\right)\,.$$

8.5 Exercises to Chapter 8

Problem 8.1:

A circular plate with radius R is simply
supported at the edge and is subjected to
a constant pressure p_0 as shown in the diagram. Using the Kirchhoff plate theory,
compute

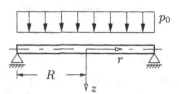

1. the deflection of the middle surface,
2. the stresses in the plate.

Problem 8.2:

The simply supported circular plate of Problem 8.1 is subjected to a singular vertical force F at the center of the plate. Using Kirchhoff's plate theory, compute the
deflection of the middle surface and the moments.

Problem 8.3:

A clamped circular plate of thickness h and radius R is heated such that after some
time a stationary temperature (difference) field $\Theta(z) = \vartheta_0 + \vartheta_1 z$ is observed across
the plate, where ϑ_0 is the temperature (difference) in the middle plane, and ϑ_1 is
the gradient of the field in z direction. Compute the stresses that are induced by this
heating.

Problem 8.4:

The clamped annular plate of Example 8.4 is subjected to a vertical shear force q_0 at
its outer edge. Compute the deflection w, the shear force q_r, and the moments m_{rr}
and $m_{\varphi\varphi}$.

Problem 8.5:

A strip of a plate with the dimensions given in the diagram below is built in at one
end and subjected to constant load (per unit length) q_R and a moment (per unit
length) m_R at the other end.

1. Calculate the maximum displacement of the plate using the Kirchhoff plate
   theory.
2. Determine the point, value, and sign of maximum stress.
3. Give the equation of the deflection curve of the plate.
4. Consider an assemblage of n beams each of width b lying next to each other as
   shown in the second diagram. Compute the deflection curve of this model and
   compare the solution with that of a plate with the width nb.

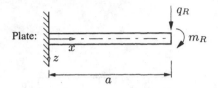

Plate:

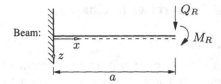

Beam:

Problem 8.6:

A spherical shell is simply supported at an angle of ϑ_0 as shown in the figure. The shell is filled with water. Compute the membrane forces for this shell.

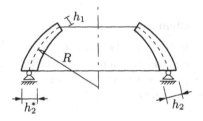

Problem 8.7:

The system shown in the figure is supposed to be a model of a soccer stadium to be constructed for the Indomitable Lions of Cameroon. If the thickness of the shell is $h(\vartheta)$, and it is loaded by its own weight only, compute the stress resultants $n_{\varphi\varphi}$ and $n_{\vartheta\vartheta}$ using the membrane theory of shells of revolution. Assume the thickness of the shell to be a linear function of ϑ.

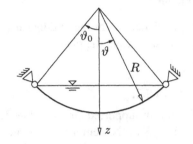

9. Stability of Equilibrium

9.1 General Remarks

In the previous chapters, we have investigated the configurations and stress distributions in systems which were in equilibrium. In this chapter, we widen the scope of our investigation by considering the behaviour of systems slightly disturbed from their equilibrium configurations. When the forces are no longer in equilibrium within a system, there will be accelerations and, in general, a complicated resulting motion. We restrict our attention here to the following question: When slightly disturbed from an equilibrium configuration, does a system tend to return to its equilibrium position or does it tend to depart even further?

For example, consider the small weight on the frictionless surfaces of Fig. 9.1. The forces on the particle are clearly in equilibrium wherever the surface is horizon-

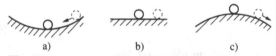

|a)|b)|c)|

Fig. 9.1 Examples of a) stable, b) neutral, and c) unstable equilibrium

tal. As the weights are slightly displaced from their equilibrium positions, different phenomena can be observed. In Fig. 9.1a, the resultant force is a restoring force; i.e., the particle is accelerated back toward the equilibrium position. Such an equilibrium is called *stable*. In Fig. 9.1c, we have the opposite situation. The resultant is an upsetting force, i.e. it accelerates the particle away from the equilibrium position. Such an equilibrium is called *unstable*. In Fig. 9.1b, we have the border line between the two previous cases; when the particle is displaced, it is again in equilibrium, and there is no tendency either to return or to go further. Such an equilibrium is said to possess *neutral stability*.

A system is said to be in a state of stable equilibrium if, for all possible geometrically admissible small displacements from the equilibrium configuration, restoring forces arise which tend to accelerate the system back toward the equilibrium position. A load-carrying structure which is in a state of unstable equilibrium is unreliable. A small disturbance can cause a cataclysmic change in configuration, and this may terminate the serviceability of the structure.

In the sequel, we are interested in those points of loading where under the same load different equilibrium configurations may exist, either under small disturbances of a deformed configuration (bifurcation problems) or related with large disturbances (snap-through problems).

9.2 Bifurcation Problems with Finite Degrees of Freedom

Let us consider a rigid, weightless bar with a pin joint at one end, and supported by a spring at the other. This bar is subject to a compressive force acting vertically in the direction of the bar axis (Fig. 9.2a). Obviously, this system is in equilibrium.

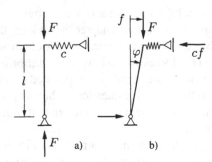

Fig. 9.2
System with one degree of freedom

Introducing now a small (admissible) disturbance f, where $f \ll l$ (Fig. 9.2b), we find

$$Ff - cfl = 0 \tag{9.1}$$

for the equilibrium of the disturbed configuration, and thus

$$(F - cl)f = 0. \tag{9.2}$$

This equation has two solutions:

1. The trivial solution $f = 0$ for the undisturbed configuration.
2. The bifurcation solution, from which the critical load is determined

$$\boxed{F_{\text{crit}} = cl.} \tag{9.3}$$

Since F is independent of deflection f it seems that the equilibrium is neutral. However, for finite disturbances $f \neq 0$, the equilibrium condition requires

$$Ff - cfl \cos \varphi = 0, \tag{9.4}$$

where

$$l \cos \varphi = \sqrt{l^2 - f^2}. \tag{9.5}$$

From this relation, we find the general condition

$$\{F - cl\sqrt{1 - (f/l)^2}\}f = 0, \tag{9.6}$$

which has the same trivial solution as before, and a second solution

$$F_{\text{crit}} = cl\sqrt{1 - (f/l)^2},$$ (9.7)

indicating an unstable equilibrium at the bifurcation point (see Fig. 9.3).

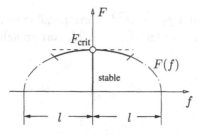

Fig. 9.3
Load-deflection curve of the rigid bar

Next, we consider a system of two pin-jointed rigid bars subjected to an axial compressive load F (see Fig. 9.4a). Again, the question is whether there can

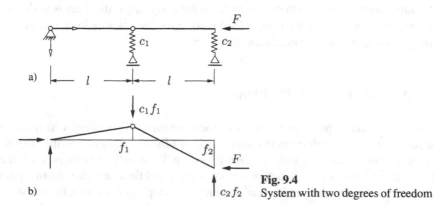

Fig. 9.4
System with two degrees of freedom

be equilibrium in the deflected form (Fig. 9.4b). As in the previous example, for small disturbances f_1, f_2, each bar is in compressive direct loading of magnitude F. Therefore, each bar exerts a force F towards the joint and for equilibrium, we find

$$F\,f_1 + (F - c_2 l)\,f_2 = 0$$
$$c_1 l f_1 + (F - 2c_2 l) f_2 = 0.$$ (9.8)

These are homogeneous equations which are satisfied by $f_1 = f_2 = 0$ corresponding to the straight form, and also when the determinant of the coefficients vanishes

$$\begin{vmatrix} F & F - c_2 l \\ c_1 l & F - 2c_2 l \end{vmatrix} = 0.$$ (9.9)

The characteristic equation

$$F^2 - F(c_1 + 2c_2)l + c_1 c_2 l^2 = 0$$ (9.10)

has the solution

$$F_{1,2} = \frac{l}{2}\left\{c_1 + 2c_2 \pm \sqrt{c_1^2 + 4c_2^2}\right\},\tag{9.11}$$

and more specifically ($c_1 = 2c_2 = c$)

$$F_{crit1} = 0.293\,cl, \quad F_{crit2} = 1.707\,cl.\tag{9.12}$$

Substituting F_{crit1} into Eqs. (9.8), we find $f_2 = \sqrt{2}f_1$ corresponding to the deflected form in Fig. 9.5a. Similarly for F_{crit2}, we find $f_2 = -\sqrt{2}f_1$ corresponding to Fig. 9.5b.

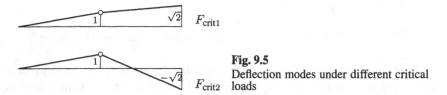

Fig. 9.5
Deflection modes under different critical loads

Summarizing, we note that these systems have as many critical loads as degrees of freedom, that the smallest critical load is the effective collapse load, and that the straight configuration is stable below this load.

9.3 A Snap-Through Problem

A different kind of buckling is the so-called snap-through buckling, which occurs in structures whose deformation under a load is of such a nature that the stiffness of the structure decreases with increasing loading. Eventually, a point is reached at which the stiffness of the structure becomes zero, and then negative. At this point the structure is unstable under dead loading and snaps into a non-adjacent stable shape.

As an example, we consider a shallow pin-jointed structure subjected to a vertical load F as shown in Fig. 9.6. Due to the compressive stresses acting in the two bars, and the shortening caused by these stresses the geometry of the structure is changed drastically so that the full nonlinearity of the geometry has to be consid-

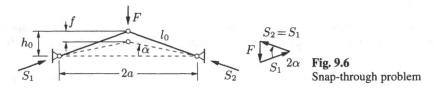

Fig. 9.6
Snap-through problem

ered. Thus, equilibrium is described in the deformed structure by

$$S_1 = S_2 = S = \frac{F}{2\sin\alpha},\tag{9.13}$$

where

$$\sin \alpha = \frac{h_0 - f}{l}, \quad l = \sqrt{a^2 + (h_0 - f)^2}.$$ (9.14)

From Hooke's law, we find

$$l = l_0 \left\{ 1 - \frac{S}{EA} \right\},$$ (9.15)

such that finally

$$F = 2EA(1 - \bar{f}) \left\{ \frac{h_0}{\sqrt{a^2 + h_0^2(1 - \bar{f})^2}} - \frac{h_0}{l_0} \right\}, \quad \bar{f} = \frac{f}{h_0}.$$ (9.16)

The load deflection behaviour of this structure is illustrated in Fig. 9.7. Portion 01 of the load-deflection curve represents the regime of constantly decreasing stiffness; point 1 is the point of zero stiffness. At that point the structure under dead loading will snap into the shape corresponding to point 2. A similar behaviour is observed for reverse loading, where now point 3 is the point of loss of stability, and the structure will snap into the shape corresponding to point 4. The critical load F_{crit} thus turns out to be the maximum (minimum) of the load-deflection curve

$$F_{\mathrm{crit}} = 2EA \frac{a}{l_0} \left\{ \left(\frac{l_0}{a} \right)^{2/3} - 1 \right\}^{3/2}$$ (9.17)

at

$$f = h_0 - a \sqrt{\left(\frac{l_0}{a} \right)^{2/3} - 1}.$$

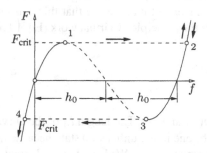

Fig. 9.7
Snap-through problem

9.4 Column Buckling

Let the originally straight column of Fig. 9.8 be in a curved position $w = w(x)$ of indifferent equilibrium under the influence of end forces F. The classical, simple Euler theory states that the bending moment in a section is Fw and that the differential equation is

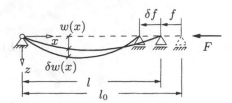

Fig. 9.8
Euler column in indifferent equilibrium

$$EJw'' + Fw = 0,\tag{9.18}$$

with the geometrical boundary conditions

$$w(0) = w(l) = 0.\tag{9.19}$$

The solution of this equation, as given in elementary textbooks, leads to the Euler buckling load or critical load

$$F_{\text{crit}} = \pi^2 \frac{EJ}{l^2}.\tag{9.20}$$

Let us now apply the principle of virtual work as stated in Chapter 1 to solve the same problem in an alternative manner. We therefore assume that the shape of the curve is $w = w(x)$, still undetermined. The bending energy stored in the beam is

$$W^* = W = \frac{1}{2} \int \frac{M^2}{EJ}\, dx = \frac{1}{2} \int EJ\,(w'')^2\, dx,\tag{9.21}$$

where as usual plane bending in the xz plane is assumed. The energy of compression

$$W^* = W = \frac{1}{2} \int \frac{N^2}{EA}\, dx = \frac{1}{2} \int EA\,(u')^2\, dx\tag{9.22}$$

has to be added to the bending energy, but when we vary the buckling deflection $w(x)$, the bending energy also varies, while the compression energy remains constant. Subjecting the beam to a virtual displacement δw, we see that this constant energy will not enter into the analysis. From the principle of virtual work (Eq. 1.85), we thus get

$$\delta W - \delta A = \int EJ\,w''\delta w''\, dx - F\delta f = 0\tag{9.23}$$

for a system in equilibrium.

We now must express δf in terms of $w(x)$ and δw. The length l_0 of the curved beam must be the same as that of the straight one in the unbuckled state because the compressive stress in the beam axis is the same for both. We have for an element

$$ds = \sqrt{dx^2 + dw^2} = dx\sqrt{1 + (w')^2},\tag{9.24}$$

and thus we find

$$l_0 = \int_0^{l_0} ds = \int_0^l \sqrt{1 + (w')^2}\, dx, \quad l_0 - l = f = \int_0^l \left\{ \sqrt{1 + (w')^2} - 1 \right\} dx.\tag{9.25}$$

The square root is approximately

$$\sqrt{1 + (w')^2} = 1 + \frac{1}{2}(w')^2 - \frac{1}{8}(w')^4 + \cdots, \tag{9.26}$$

so that, neglecting powers higher than 2 of the small slope,

$$f = \frac{1}{2}\int_0^l (w')^2 \, dx, \quad \rightarrow \quad \delta f = \int_0^l w' \, \delta w' \, dx. \tag{9.27}$$

Thus, the solution of the foregoing buckling problem is reduced to finding the stationary value (minimum) of the potential energy Π

$$\Pi = W - A = \frac{1}{2}\int_0^l EJ\,(w'')^2 \, dx - \frac{1}{2}\,F\int_0^l (w')^2 \, dx. \tag{9.28}$$

We note here that the general form of the stored energy W also contains contributions of springs which may be added to the structure at discrete points, e.g. as supports, the so-called Dirichlet boundary terms,

$$W = \frac{1}{2}\int_0^l EJ\,(w'')^2 \, dx + \frac{1}{2}\sum_i c\,w^2(x_i) + \frac{1}{2}\sum_j c^*(w')^2(x_j), \tag{9.29}$$

where c, and c^*, respectively, are the extensional and bending stiffnesses of the springs.

The above principle (Eq. 9.28) can now be used to determine approximate solutions. Let $\tilde{w}(x)$ be a reasonable, approximate function which for a differential equation of order $2m$ is m times continuously differentiable and which satisfies at least the geometrical boundary conditions. We will call this an *admissible function*. It is observed then that the principle is equivalent to finding, among all admissible functions $\tilde{w}$, those which make the quotient

$$F_{\text{crit}} \le R[\tilde{w}] = \frac{\displaystyle\int EJ\,(\tilde{w}'')^2 \, dx}{\displaystyle\int (\tilde{w}')^2 \, dx} \tag{9.30}$$

stationary. This is Rayleigh's method for computing an approximation to the lowest buckling load. The approximation obtained by this method is never less than the true value of F_{crit}.

We can improve this method by introducing admissible functions, e.g.

$$\tilde{w}(x) = \sum_{i=1}^r a_i \tilde{w}_i(x), \tag{9.31}$$

which are still functions of r undetermined parameters a_i, whose relative values are to be selected, in conjunction with $R[\tilde{w}_i(x; a_i)]$, such as to make Π stationary. The necessary and sufficient conditions for $\delta\Pi$ to vanish are

$$\frac{\partial \Pi}{\partial a_i} = 0, \quad (i = 1, 2, \ldots, r), \tag{9.32}$$

representing r linear homogeneous equations in the a_i

$$\boxed{\{\mathbf{L} - \tilde{F}\mathbf{M}\}\mathbf{a} = 0\,,}$$ (9.33)

wherein

$$L_{ik} = \int_0^l EJ\,\tilde{w}_i''\tilde{w}_k''\,\mathrm{d}x + c\tilde{w}_i(x_0)\tilde{w}_k(x_0) + c^*\tilde{w}_i'(x_1)\tilde{w}_k'(x_1)$$

(9.34)

$$M_{ik} = \int_0^l \tilde{w}_i'\tilde{w}_k'\,\mathrm{d}x\,.$$

These equations have solutions other than the trivial one only if the determinant of the coefficients of the a_i vanishes

$$\det(\mathbf{L} - \tilde{F}\mathbf{M}) = 0\,,$$ (9.35)

yielding a sequence of r approximate buckling loads. Each of these approximate critical loads is never less than the corresponding one of the first r true critical loads. This is the so-called Rayleigh-Ritz method for computing approximations to the r lowest eigenvalues.

Example 9.1:

Determine the critical load of the Euler column, case 3, as shown in the figure. The beam is fixed at one end, while the other end is simply supported and under a compressive load F.

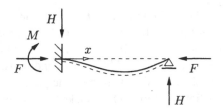

Solution:

The geometrical boundary conditions of the problem are

$$w(0) = w'(0) = w(l) = 0\,.$$

1. The simplest admissible function $\tilde{w}(x)$ fulfilling these conditions would be

$$\tilde{w}(x) = cx^2(x - l)\,,$$

where c is a constant coefficient which will play no role in the solution. To substitute this shape into Eq. (9.30), we need

$$\tilde{w}'(x) = c(3x^2 - 2xl)\,,$$

$$\tilde{w}''(x) = c(6x - 2l)\,,$$

and

$$\int_0^l EJ\,(\tilde{w}'')^2\,\mathrm{d}x = 4c^2EJ\,l^3\,, \quad \int_0^l (\tilde{w}')^2\,\mathrm{d}x = \frac{2}{15}c^2l^5\,.$$

Thus, we find as approximation for the buckling load

$$F_{crit} \leq R[\tilde{w}] = 30 \frac{EJ}{l^2}.$$

This result is considerably greater than the correct value of $20.19\ EJ/l^2$.

2. A better result is obtained by using the shape

$$\tilde{w}(x) = cx^2\left(x^2 - \frac{5}{2}xl + \frac{3}{2}l^2\right),$$

which in addition to satisfying the three geometrical conditions above also satisfies the dynamical condition

$$M(l) = -EJ\,w''(l) = 0.$$

A function satisfying all boundary conditions will be called a *test function*. We further note that test functions are obtained e.g. from any deflection curve of the system under consideration, since they satisfy all boundary conditions. Here, we took the deflection curve of the beam subject to a constant distributed lateral load q_0. Introducing this function into the Rayleigh-quotient Eq. (9.30), we arrive at

$$F_{crit} \leq R[\tilde{w}] = 21 \frac{EJ}{l^2}.$$

3. The function

$$\tilde{w}(x) = c(\cos\xi - \cos 3\xi), \quad \xi = \pi x/2l,$$

is an especially good representation of the buckling mode, leading to the result

$$F_{crit} \leq R[\tilde{w}] = 20.23 \frac{EJ}{l^2}.$$

This exceeds the correct value by only 0.2 %.

4. To obtain a Rayleigh-Ritz solution, we introduce an approximation

$$\tilde{w}(x) = c_1 w_1(x) + c_2 w_2(x)$$

of two linearly independent admissible shape functions

$$w_1(x) = x^2(x-l), \quad w_2(x) = x^3(x-l),$$

with derivatives

$$w_1'(x) = 3x^2 - 2xl, \quad w_2'(x) = 4x^3 - 3x^2l,$$

$$w_1''(x) = 6x - 2l, \quad w_2''(x) = 12x^2 - 6xl.$$

From these, we find

$$L_{11} = \int_0^l EJ\,(w_1'')^2\,\mathrm{d}x = 4EJ\,l^3$$

$$L_{12} = \int_0^l EJ\,w_1''w_2''\,\mathrm{d}x = 4EJ\,l^4$$

$$L_{22} = \int_0^l EJ\,(w_2'')^2\,\mathrm{d}x = \frac{24}{5}EJ\,l^5$$

and

$$M_{11} = \int_0^l EJ\,(w_1')^2\,\mathrm{d}x = \frac{2}{15}\,EJ\,l^5$$

$$M_{12} = \int_0^l EJ\,w_1'\,w_2'\,\mathrm{d}x = \frac{1}{10}\,EJ\,l^6$$

$$M_{22} = \int_0^l EJ\,(w_2')^2\,\mathrm{d}x = \frac{3}{35}\,EJ\,l^7$$

Introducing these values into Eq. (9.35), gives

$$\begin{vmatrix} 1 - \dfrac{2}{3}\lambda & 1 - \dfrac{1}{2}\lambda \\[2mm] 1 - \dfrac{1}{2}\lambda & \dfrac{6}{5} - \dfrac{3}{7}\lambda \end{vmatrix} = 0,$$

where

$$\lambda = \frac{\tilde{F}l^2}{20\,EJ}.$$

The characteristic equation of this problem

$$\lambda^2 - \frac{32}{5}\lambda + \frac{28}{5} = 0$$

has the two solutions

$$\lambda_{1,2} = \frac{16}{5} \pm \frac{1}{5}\sqrt{116},$$

and thus

$$\tilde{F}_1 = 20.92\ EJ/l^2, \quad (\text{exact:}\quad 20.19\ EJ/l^2),$$

$$\tilde{F}_2 = 107.1\ EJ/l^2, \quad (\text{exact:}\quad 59.69\ EJ/l^2).$$

9.5 Exercises to Chapter 9

Problem 9.1:

1. Using the quotient

$$F_{\text{crit}} \le R[\tilde{w}] = \frac{-\int\limits_0^l EJ\tilde{w}''\tilde{w}\,\mathrm{d}x}{\int\limits_0^l \tilde{w}^2\,\mathrm{d}x}$$

with the admissible function $\tilde{w} = c\xi(1 - \xi)$, $\xi = x/l$,

2. and the energy method, i.e. the Rayleigh quotient

$$F_{crit} \leq R[\tilde{w}] = \frac{\int_0^l (EJ\tilde{w}'')^2 \, dx}{\int_0^l (\tilde{w}')^2 \, dx}$$

with the same function $\tilde{w} = c\,\xi(1-\xi)$,

determine approximate solutions for the critical load F_{crit} for a column that is simply supported at one end and pinned at the other (Euler column, second case). Compare these results with the exact solution.

Problem 9.2:

For the system shown in the figure, determine approximate solutions for the critical load F_{crit}

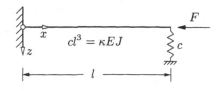

1. using the Rayleigh quotient with the admissible function $\tilde{w} = a\,\xi^2$ and
2. the Rayleigh-Ritz method with the functions $\tilde{w}_1 = \xi^2$ and $\tilde{w}_2 = \xi^3$,

where $\xi = x/l$. By setting $c \to 0$ and $c \to \infty$, compare the results obtained with the corresponding exact solutions.

Problem 9.3:

For the system shown in the figure, determine from the Rayleigh quotient approximate solutions for the critical load F_{crit} using

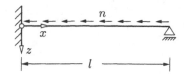

1. as admissible function the polynomial $\tilde{w} = a\,\xi(1-\xi)$, $\xi = x/l$,
2. and by using the Rayleigh-Ritz method with the functions $\tilde{w}_1 = \xi(1-\xi)$ and $\tilde{w}_2 = \xi^3(1-\xi)$.

Problem 9.4:

For a column that is fixed at one end, pinned at the other end and loaded by a constant normal force per unit length of intensity n as shown in the diagram, compute

1. the Rayleigh quotient for the normal-force n.

2. From this quotient, calculate using the function $\tilde{w} = a\,\xi^2(1 - \xi)$, $\xi = x/l$, an approximate solution for the critical load n_{crit}.

3. Determine a polynomial of degree 4 that satisfies all the boundary conditions and use this polynomial with the Rayleigh quotient from 1. to compute an approximate solution for the critical load n_{crit}.

10. Some Basic Concepts of Dynamics

10.1 Principle of Virtual Work

In the following last chapters, we will again widen the scope of our investigations by considering the motion of particles, masses and bodies, and the forces acting on these moving bodies. In Chapter 1, we discussed the equations of motion and the balance of energy of bodies in motion (Sections 1.5 and 1.6), and we introduced the principle of virtual work for statical problems (Section 1.7). We shall now extend this principle from statics to dynamics by adding to the equations of motion - on the right-hand side - the acceleration (inertia) terms, which have been disregarded when determining Eq. (1.85).

As in Section 1.7, we introduce a virtual displacement δu_i, and mention that for dynamical problems in addition to the conditions stated there, virtual displacements have to be performed at a fixed time.

Again, starting by multiplying the equation of motion (Eq. 1.63) by δu_k, we arrive at

$$\int_V \left\{ \frac{\partial \sigma_{ik}}{\partial x_i} + \rho\, b_k \right\} \delta u_k \, \mathrm{d}V = \int_V \rho\, a_k \delta u_k \, \mathrm{d}V. \tag{10.1}$$

The right-hand side now gives

$$\int_V \rho\, \dot{v}_k \delta u_k \, \mathrm{d}V = \frac{\mathrm{D}}{\mathrm{d}t} \int_V \rho\, v_k \delta u_k \, \mathrm{d}V - \int_V \rho\, v_k \frac{\mathrm{D}}{\mathrm{d}t} (\delta u_k) \, \mathrm{d}V. \tag{10.2}$$

Since $\delta(\mathrm{D}u_i) = \mathrm{D}(\delta u_i)$, the second term on the right side can be written as the variation of the kinetic energy E

$$\int_V \rho\, v_k \frac{\mathrm{D}}{\mathrm{d}t} (\delta u_k) \, \mathrm{d}V = \int_V \rho\, v_k \delta v_k \, \mathrm{d}V = \delta \int_V \frac{1}{2} \rho\, v_k^2 \, \mathrm{d}V = \delta E. \tag{10.3}$$

Thus, with the result of Section 1.7 in mind, we find

$$-\delta W + \delta A + \delta E - \frac{\mathrm{D}}{\mathrm{d}t} \int_V \rho\, v_k \delta u_k \, \mathrm{d}V = 0, \tag{10.4}$$

the principle of virtual work for dynamical problems. We note that this principle, too, is independent of any constitutive relation.

10.2 Hamilton's Principle

We will now integrate Eq. (10.4) with respect to the time t between two definite time limits t_1 and t_2, which may be chosen arbitrary. The integration of the last term gives another boundary term. We wish, however, that the variations δu_k vanish at the two time limits t_1 and t_2, and call that a variation between definite limits. In this case, the boundary term drops out, and we obtain Hamilton's principle

$$\int_{t_1}^{t_2} (-\delta W + \delta A + \delta E)\, dt = 0. \tag{10.5}$$

Introducing now an elastic material, where the stresses are derivable from a strain energy potential $w(\varepsilon_{ik})$, and assuming moreover that the forces s_k and b_k are also derivable from a potential, we can introduce a single scalar function

$$\Pi = W + \Phi = \int_V \rho\, w(\varepsilon_{ik})\, dV + \int_V \rho\, \varphi_V\, dV + \int_A \varphi_A\, dA, \tag{10.6}$$

where

$$b_k = -\frac{\partial \varphi_V}{\partial u_k}, \quad s_k = -\frac{\partial \varphi_A}{\partial u_k}. \tag{10.7}$$

A system of this kind is called a conservative system with $\delta\Phi = -\delta A$. Since we have

$$\delta\Pi = \delta W - \delta A, \tag{10.8}$$

Hamilton's principle, for the case of conservative forces, may be written as follows

$$\boxed{\delta \int_{t_1}^{t_2} (E - \Pi)\, dt = \delta \int_{t_1}^{t_2} L\, dt = 0,} \tag{10.9}$$

where $L = E - \Pi$ is called the Lagrangian function.

This principle is very similar in its character to the principles discussed so far. Once more, we have a stationary principle, where now the quantity which has to be made stationary is the time integral of the Lagrangian function L.

10.3 The Euler-Lagrange Equations

We now restrict our considerations to the motion of rigid bodies ($\delta W = 0$), and consider an ensemble of coupled objects. The position of this system at any instant is called its configuration, and we call the set of parameters needed to specify the configuration the generalized coordinates q_i ($i = 1, 2, \ldots, \lambda$). The number λ of independent parameters determines the degree of freedom of the system.

Moreover, we introduce dual generalized forces Q_i as (work) conjugated pairs of the q_i, such that they act in the direction of the q_i, and

$$\delta A = Q_i \delta q_i \tag{10.10}$$

is the virtual work done by these forces in the direction of the virtual displacements δq_i

The essential characteristic of a system is that the objects are coupled, and hence restricted or constrained in their motion. If we consider, e. g. an ensemble of n rigid bodies, we have the degree of freedom

$$\lambda = 6n - r\,, \tag{10.11}$$

where r are the number of constraints. These constraints may have the form

$$\boxed{f_k(q_i, t) = 0\,, \quad i = 1, 2, \ldots, \lambda + r\,; \quad k = 1, 2, \ldots, r} \tag{10.12}$$

of holonomic rheonomic functions if they explicitly depend on time, i.e. if the constraints are moving. If the constraints are fixed, they are called scleronomic.

We note in passing that constraints which are given in the form of non-integrable differential conditions are called non-holonomic. In the sequel, we will restrict our consideration to holonomic systems. Finding the minimum of the definite integral of Eq. (10.9) can be solved by starting from the Lagrangian function L as the difference between the kinetic and the potential energy. We will leave the definition of L free, except for the general assumption that it is some given function of the coordinates q_i and their time derivatives $\dot{q}_i$; for the sake of generality, we will even allow an explicit dependence of L on the time t

$$L = L(q_1, \ldots, q_i\,; \dot{q}_1, \ldots, \dot{q}_i\,; t)\,, \quad i = 1, \ldots, \lambda\,. \tag{10.13}$$

Following Hamilton's principle (Eq. 10.9), we need

$$\delta L = \frac{\partial L}{\partial q_i}\delta q_i + \frac{\partial L}{\partial \dot{q}_i}\delta\dot{q}_i\,. \tag{10.14}$$

Using

$$\delta\dot{q}_i = \frac{\mathrm{D}}{\mathrm{d}t}(\delta q_i)\,, \tag{10.15}$$

the second term on the right-hand side of this relation can be transformed such that instead of Eq. (10.14), we have

$$\delta L = \frac{\partial L}{\partial q_i}\delta q_i + \frac{\mathrm{D}}{\mathrm{d}t}\left(\frac{\partial L}{\partial \dot{q}_i}\delta q_i\right) - \frac{\mathrm{D}}{\mathrm{d}t}\left(\frac{\partial L}{\partial \dot{q}_i}\right)\delta q_i\,. \tag{10.16}$$

Introducing this expression into Eq. (10.9) yields

$$\int_{t_1}^{t_2}\delta L\,\mathrm{d}t = \left[\frac{\partial L}{\partial \dot{q}_i}\delta q_i\right]_{t_1}^{t_2} - \int_{t_1}^{t_2}\left\{\frac{\partial L}{\partial q_i} - \frac{\mathrm{D}}{\mathrm{d}t}\left(\frac{\partial L}{\partial \dot{q}_i}\right)\right\}\delta q_i\,\mathrm{d}t = 0\,. \tag{10.17}$$

The first term on the right side drops out, since all variations are supposed to vanish at the two end points of integration. The remaining term has to vanish, too, and thus since the δq_i are independent, this leads to λ simultaneous equations

$$\frac{\mathrm{D}}{\mathrm{d}t}\left(\frac{\partial L}{\partial \dot{q}_i}\right) - \frac{\partial L}{\partial q_i} = 0, \quad (i = 1, 2, \ldots, \lambda). \tag{10.18}$$

These are the Euler-Langrange equations for conservative systems.

Since the potential energy $\Phi = \Phi(q_i; t)$ is no function of generalized velocities, we also can express Eq. (10.18) through the kinetic energy and the generalized forces

$$\frac{\mathrm{D}}{\mathrm{d}t}\left(\frac{\partial E}{\partial \dot{q}_i}\right) - \frac{\partial E}{\partial q_i} = -\frac{\partial \Phi}{\partial q_i} = Q_i, \quad (i = 1, 2, \ldots, \lambda). \tag{10.19}$$

For non-conservative forces, we start from Eq. (10.5), and then using Eq. (10.10), we realize that the second part of Eq. (10.19) is also valid if the Q_i are not derivable from a potential

$$\frac{\mathrm{D}}{\mathrm{d}t}\left(\frac{\partial E}{\partial \dot{q}_i}\right) - \frac{\partial E}{\partial q_i} = Q_i, \quad (i = 1, 2, \ldots, \lambda). \tag{10.20}$$

We finally note that Eqs. (10.18) and (10.20) are also valid for systems of deformable bodies.

10.4 The Lagrangian Multiplier Method

Given the Lagrangian function L expressed as function of $\lambda + r$ variables and the time, and given r holonomic auxiliary conditions of the kind (Eq. 10.12), it is sometimes tedious to eliminate the redundant r variables from the Lagrangian.

In this case, following a suggestion of Lagrange, we simply modify L by adding the auxiliary conditions

$$L' = L + \lambda_k f_k, \quad (k = 1, 2, \ldots, r), \tag{10.21}$$

where the Lagrange multipliers λ_k are r additional variables. Introducing now this modified Lagrangian into the variational principle gives the following $\lambda + r$ equations of motion

$$\frac{\mathrm{D}}{\mathrm{d}t}\left(\frac{\partial E}{\partial \dot{q}_i}\right) - \frac{\partial E}{\partial q_i} = Q_i + \lambda_k \frac{\partial f_k}{\partial q_i}, \quad (i = 1, 2, \ldots, \lambda + r). \tag{10.22}$$

Together with the r auxiliary conditions, we thus have $\lambda + 2r$ unknowns q_i and λ_k. The merit of this multiplier method is that now the equations of motion, together with the given auxiliary conditions, appear as the consequence of one single and unrestricted variational principle.

11. Oscillators With One Degree of Freedom

11.1 Undamped Free Vibration

The simplest oscillatory system is one of a single degree of freedom, the motion of which can be described by a single coordinate q. A mass m suspended from a spring of negligible mass represents such a system.

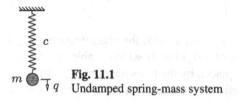

Fig. 11.1
Undamped spring-mass system

We take as reference the statical equilibrium position of the mass. In this position, the gravitational force $Q_{st} = mg$ acting on the mass is balanced by the spring force cq_{st}, where c is the spring stiffness, defined as the force required per unit extension or compression, and q_{st} is the statical deflection of the spring due to the weight mg. It is evident then that by choosing the origin of the coordinate q at the static equilibrium position, only forces due to displacement from this position should be considered. From Newton's second law, or alternatively using d'Alembert's principle, we find

$$m\ddot{q} + cq = 0, \tag{11.1}$$

and

$$\boxed{\ddot{q} + \omega_0^2 q = 0,} \tag{11.2}$$

having introduced the circular (or eigen-) frequency

$$\omega_0 = \sqrt{\frac{c}{m}}. \tag{11.3}$$

The homogeneous second-order differential equation has the general solution

$$q(t) = c_0 \sin \omega_0 t + c_1 \cos \omega_0 t, \tag{11.4}$$

describing a motion with the period

$$T = \frac{2\pi}{\omega_0} = 2\pi \sqrt{\frac{m}{c}}, \tag{11.5}$$

and frequency

$$f = T^{-1} = \frac{1}{2\pi} \sqrt{\frac{c}{m}}. \tag{11.6}$$

Equation (11.6) may be expressed in terms of the statical deflection q_{st} by noting that $cq_{st} = mg$, from which we obtain

$$f = \frac{1}{2\pi} \sqrt{\frac{g}{q_{st}}}. \tag{11.7}$$

Thus the frequency is found to be a function of the statical deflection of a system.

The constants c_o and c_1 in Eq. (11.4) are obtained from the initial conditions. In the most general case, the system may be started from position $q = q_o$ with velocity $\dot{q} = \dot{q}_o$

$$q(t) = \frac{\dot{q}_o}{\omega_0} \sin \omega_0 t + q_0 \cos \omega_0 t. \tag{11.8}$$

The differential equation (11.2) and its solution with the eigenfrequency ω_o may also be applied to a system consisting of a rigid body which is able to rotate about a fixed axis. In this case m must be replaced by the mass moment of inertia θ of the body, and the spring stiffness by c^* defined as the moment per unit angle of rotation.

For a system of mass m, the mass moments of inertia θ_{ik} are defined analogously to Eq. (3.22) as the integrals

$$\theta_{ik} = \begin{pmatrix} \theta_{xx} & \theta_{xy} & \theta_{xz} \\ \theta_{yx} & \theta_{yy} & \theta_{yz} \\ \theta_{zx} & \theta_{zy} & \theta_{zz} \end{pmatrix} = \int_V \{(x_r x_r)\delta_{ik} - x_i x_k\}\rho \, dV, \quad x_i = x, y, z, \tag{11.9}$$

where the x_i are centroidal axes, and the integrals are taken over the whole mass or volume, respectively, of the system.

We note that due to their definition the mass moments of inertia θ_{ik} - just as the moments of inertia J_{ik} - can be interpreted as components of a second-rank tensor, and as such have the same properties and follow the same rules as the latter.

The single degree-of-freedom (DOF) system is an idealization which in many cases is justified. Such approximations give excellent results when the mass of the elastic element is small compared to the lumped mass which it supports. The eigenfrequency of the system can then be determined with good accuracy from Eq. (11.3), provided the proper stiffness c is used. For complex systems of one degree of freedom, the total stiffness is expressed as function of the single stiffnesses c_i via

$$c = \sum_i c_i \quad \text{for a parallel connection}$$

$$\frac{1}{c} = \sum_i \frac{1}{c_i} \quad \text{for a series connection.} \tag{11.10}$$

The stiffness of elastic parts of the system is determined from the influence coefficients δ_{ik} of elastostatics (refer to Section 6.6). The deflection q_i at a point i of a system under load X_i acting at this point is

$$q_i = \delta_{ii} X_i, \quad \sum i. \tag{11.11}$$

where δ_{ii} is the influence coefficient, i.e. the deflection under a unit load $X_i = 1$. Thus, from the inverse relation $X_i = c\,q_i$, we conclude for a single DOF system

$$
\begin{aligned}
c &= \frac{1}{\delta_{11}} \quad \rightarrow \quad \omega_0^2 = \frac{c}{m} = \frac{1}{m\delta_{11}}, \quad \text{and} \\
c^* &= \frac{1}{\delta_{11}} \quad \rightarrow \quad \omega_0^2 = \frac{c^*}{\theta} = \frac{1}{\theta\delta_{11}}.
\end{aligned}
\tag{11.12}
$$

Moreover, in cases where the system contains elastic parts - like beams etc. - it is also advantageous to determine the differential equations of the system using methods from statics. We will demonstrate this with a simple cantilever beam with one concentrated mass at its end.

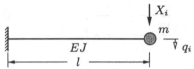

Fig. 11.2
Cantilever beam with one concentrated mass

Starting with equation (11.11), we know from d'Alembert's principle the magnitude of X to describe the inertia

$$X_i = -m\ddot{q}_i. \tag{11.13}$$

Thus, we find

$$q_i = -\delta_{ii}\, m\ddot{q}_i. \tag{11.14}$$

Multiplying this result with the inverse of δ_{ii}, we arrive at

$$\boxed{m\ddot{q}_i + c\,q_i = 0,} \tag{11.15}$$

the differential equation of free vibration of the mass m at point i.

This method can also be applied to systems of several DOF. From statics, we know

$$
\begin{aligned}
q_1 &= \delta_{11} X_1 + \delta_{12} X_2 + \cdots + \delta_{1n} X_n \\
q_2 &= \delta_{21} X_1 + \delta_{22} X_2 + \cdots + \delta_{2n} X_n \\
&\cdots \\
q_n &= \delta_{n1} X_1 + \delta_{n2} X_2 + \cdots + \delta_{nn} X_n
\end{aligned}
\tag{11.16}
$$

for a system of n DOF (refer to Eq. 6.52), i.e.

$$\mathbf{q} = \boldsymbol{\delta}\mathbf{X}. \tag{11.17}$$

Herein δ is the matrix of the influence coefficients, and according to d'Alembert's principle $\mathbf{X}$ is the vector of the inertia forces: $-m_i \ddot{q}_i$, and $-\theta_i \ddot{q}_i$, respectively, or in matrix notation

$$\mathbf{X} = -\mathbf{M}\ddot{\mathbf{q}}. \tag{11.18}$$

Introducing this into Eq. (10.16), we finally arrive

$$\mathbf{q} = -\boldsymbol{\delta}\,\mathbf{M}\ddot{\mathbf{q}}. \tag{11.19}$$

Multiplying this relation from the left-hand side with $\boldsymbol{\delta}^{-1} = \mathbf{C}$ gives

$$\boxed{\mathbf{M}\ddot{\mathbf{q}} + \mathbf{C}\mathbf{q} = 0,} \tag{11.20}$$

the system of differential equations describing free vibrations.

Example 11.1:

Compute the natural frequency of the system shown in the figure. The spring has a stiffness of $c = \alpha EJ/l^3$.

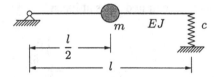

Solution:

We refer to Eq. (11.12) and determine the influence coefficient δ_{11} of the system for a unit load $X_1 = 1$ acting vertically at the position of the concentrated mass m. The moment diagram turns out to be (refer to Section 6.6)

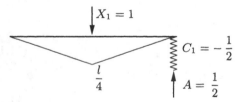

and thus we calculate

$$EJ\delta_{11} = \frac{l^3}{48} + \frac{1}{4}\frac{EJ}{c} = \frac{l^3}{48}\left(1 + \frac{12}{\alpha}\right),$$

$$\omega_0^2 = \frac{48EJ}{ml^3}\frac{\alpha}{\alpha + 12},$$

with bounds $0 \le \omega_0^2 \le \dfrac{48EJ}{ml^3}$ for the free, and the simply supported beam, respectively. Taking $\alpha = 12$ (refer to Example 6.2), we get

$$\omega_0^2 = \frac{24EJ}{ml^3}.$$

11.2 Damped Free Vibration

Energy dissipation takes place in all vibrating systems, and the force associated with the dissipation is called damping force. There are different types of damping forces, and the one which is the simplest to consider is a viscous damping proportional to the velocity.

Fig. 11.3
Damped system of one degree of freedom

The equation of motion for the damped free vibration is

$$m\ddot{q} + d\dot{q} + cq = 0,$$

(11.21)

or

$$\ddot{q} + 2D\omega_0\dot{q} + \omega_0^2 q = 0,$$

(11.22)

where

$$D = \frac{d}{2m\omega_0}$$

(11.23)

is the (dimensionless) damping factor.

Equation (11.22) is a homogeneous second-order differential equation which can be solved by assuming a (complex) function of the form

$$q(t) = \hat{q}e^{st}.$$

(11.24)

Upon substitution, we find that s must satisfy the characteristic equation

$$F(s) = s^2 + 2D\omega_0 s + \omega_0^2 = 0,$$

(11.25)

which has the two roots

$$s_{1,2} = \omega_0\left(-D \pm \sqrt{D^2 - 1}\right).$$

(11.26)

Hence, the general solution becomes

$$q(t) = c_0 e^{s_1 t} + c_1 e^{s_2 t},$$

(11.27)

where c_o and c_1 must be evaluated from initial conditions.

The free motion of the damped system depends on the numerical value of the radical of Eq. (11.26). As a reference quantity we define critical damping as the value of D that reduces this radical to zero, i.e. $D_c^2 = 1$. The actual damping of the system can then be specified by the damping factor:

Case 1, $D > 1$: The radical in this case is real, and is always less then D, so that

s_1 and s_2 are negative. The displacement q then becomes the sum of two decaying exponentials (Eq. 11.27), with

$$c_0 = \frac{\dot{q}_0 + q_0 \omega_0 \left(D + \sqrt{D^2 - 1}\right)}{2\omega_0 \sqrt{D^2 - 1}},$$

$$c_1 = -\frac{\dot{q}_0 + q_0 \omega_0 \left(D - \sqrt{D^2 - 1}\right)}{2\omega_0 \sqrt{D^2 - 1}}.$$

(11.28)

Such a motion is non-oscillatory and is called aperiodic.

Case 2, $D < 1$: The radical in this case is imaginary, and s can be written as

$$s_{1,2} = -\delta \pm i\omega,$$

(11.29)

where

$$\omega = \omega_0 \sqrt{1 - D^2}$$

(11.30)

is the circular frequency of the damped system. The general solution then becomes

$$\boxed{q(t) = e^{-D\omega_0 t} \left(c_0 \sin \omega t + c_1 \cos \omega t\right),}$$

(11.31)

and the motion is oscillatory with decreasing amplitude, with

$$c_0 = \frac{\dot{q}_0 + D q_0 \omega_0}{\omega}, \quad c_1 = q_0.$$

(11.32)

Case 3, $D = 1$: This case (critical damping) represents a transition between oscillatory and non-oscillatory conditions. The roots s_1 and s_2 approach each other, and become equal to $-\omega_0$. The general solution then has the form

$$\boxed{q(t) = e^{-\omega_0 t}[q_0(1 + \omega_0 t) + \dot{q}_0 t].}$$

(11.33)

The motion is similar to the aperiodic motion of case 1, however, it has the smallest damping possible for aperiodic motion, and hence returns to the equilibrium position in the shortest time.

The amount of damping present in an oscillatory system can be determined by measuring the rate of decay of oscillation. For this purpose, we define the logarithmic decrement, which is the natural logarithm of the ratio of any two successive maxima q_i, q_{i+1} of $q(t)$ as

$$\vartheta = \ln \frac{q_i}{q_{i+1}} = \ln \frac{q(t)}{q(t + T)}.$$

(11.34)

If we let the damped oscillation be described by Eq. (11.31), the logarithmic decrement defined by Eq. (11.34) can be determined in terms of the damping factor

$$\vartheta = \ln \frac{e^{-D\omega_0 t}}{e^{-D\omega_0 (t+T)}} = D\omega_0 T.$$

(11.35)

Since the period of a damped oscillation is equal to

$$T = \frac{2\pi}{\omega} = \frac{2\pi}{\omega_0 \sqrt{1 - D^2}},$$

(11.36)

the logarithmic decrement in terms of D becomes

$$\vartheta = 2\pi \, \frac{D}{\sqrt{1 - D^2}} \, , \tag{11.37}$$

which can be approximated for small damping factors as

$$\vartheta \cong 2\pi D \, . \tag{11.38}$$

This allows the determination of D from a number of cycles of a simple oscillation test.

We finally mention that any harmonic oscillation can also be described through

$$q(t) = \hat{q} \, \cos \left(\omega t + \varphi \right), \tag{11.39}$$

where now φ is some phase angle, and $\hat{q}$ is the amplitude. Compared with Eq. (11.8), we find

$$
\begin{aligned}
q_0 &= \hat{q} \cos \varphi, \qquad \dot{q}_0/\omega = -\hat{q} \sin \varphi \\
\hat{q} &= \sqrt{q_0^2 + \frac{\dot{q}_0^2}{\omega^2}} \, , \qquad \varphi = \arctan \frac{-\dot{q}_0}{\omega q_0} \, .
\end{aligned}
\tag{11.40}$$

11.3 Forced Vibration with Harmonic Excitation

We consider here a viscously damped spring-mass system (see Fig. 11.3) excited by a harmonic force $Q(t) = \hat{Q} \cos \Omega t$. The differential equation of motion can then be written as

$$\boxed{\ddot{q} + 2 D \omega_0 \, \dot{q} + \omega_0^2 \, q = \frac{\hat{Q}}{m} \, \cos \Omega t \, ,} \tag{11.41}$$

where Ω is the frequency of harmonic excitation.

The solution can be considered in two parts. There will be some damped free vibration at its eigenfrequency, which is the solution of the homogeneous equation, and a particular solution, which is a steady oscillation at the frequency of excitation. The free vibration is introduced as a transient by the condition at $t = 0$ and will gradually disappear because of damping. The steady-state oscillation therefore will be described by the particular solution

$$\boxed{q(t) = \hat{q} \, \cos(\Omega t + \varphi) \, ,} \tag{11.42}$$

with

$$
\begin{aligned}
\hat{q}(\eta) &= \frac{\hat{Q}/c}{\sqrt{(1 - \eta^2)^2 + 4 D^2 \eta^2}} \, , \\
\varphi(\eta) &= \arctan \frac{-2 D \eta}{1 - \eta^2} \, ,
\end{aligned}
\tag{11.43}$$

where $\eta = \Omega/\omega_o$. For convenience of presentation and discussion, this solution is generally reduced to non-dimensional form. We therefore introduce the magnification factor of amplitudes

$$V_a(\eta) = \frac{1}{\sqrt{(1-\eta^2)^2 + 4D^2\eta^2}}, \qquad (11.44)$$

relating the amplitude of the response to that of harmonic excitation, and the related phase response

$$\psi_a(\eta) = -\varphi. \qquad (11.45)$$

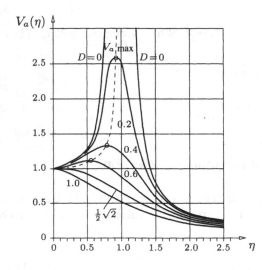

Fig. 11.4
Magnification factor of amplitudes

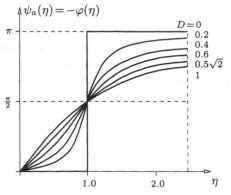

Fig. 11.5
Phase response

The maximum of V_a is

$$V_{a\,max} = \frac{1}{2\,D\sqrt{1-D^2}} \quad \text{at} \quad \eta = \sqrt{1-2D^2} \qquad (11.46)$$

for damping factors of $0 < D \le \frac{1}{2}\sqrt{2}$.

A vibratory system is sometimes excited by a prescribed motion of some point in the system. If we let $u(t)$ be the harmonic motion of the support point of the system of Fig. 11.6,

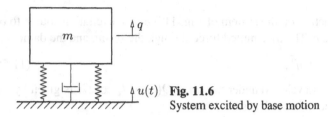

Fig. 11.6
System excited by base motion

the differential equation becomes

$$m\ddot{q} + d(\dot{q} - \dot{u}) + c(q - u) = 0,$$ (11.47)

which may be rearranged to

$$\ddot{q} + 2D\omega_0\dot{q} + \omega_0^2 q = \omega_0^2 u + 2D\omega_0\dot{u}.$$ (11.48)

We will introduce now the method of complex algebra, which often simplifies the task of determining the steady-state solution. We have

$$u(t) = \hat{u}e^{i\Omega t}, \quad q(t) = \hat{q}e^{i(\Omega t+\varphi)}.$$ (11.49)

Substituting these into the differential equation, we obtain

$$(1 + 2iD\eta - \eta^2)\hat{q}e^{i\varphi} = (1 + 2iD\eta)\hat{u},$$ (11.50)

from which the (complex) amplitude ratio is

$$Y = \frac{1 - \eta^2 + 4D^2\eta^2 - 2iD\eta^3}{(1 - \eta^2)^2 + 4D^2\eta^2}.$$ (11.51)

The absolute value of Y is then

$$|Y| = \sqrt{Re(Y)^2 + Im(Y)^2} = \sqrt{1 + 4D^2\eta^2}\, V_a(\eta),$$ (11.52)

the magnification factor of amplitudes of the prescribed displacement.
    To find the phase angle φ, we put

$$e^{i\varphi} = \cos\varphi + i\sin\varphi,$$ (11.53)

and again find from Eq. (11.50)

$$\tan\varphi = \frac{Im(Y)}{Re(Y)} = -\frac{2D\eta^3}{1 - \eta^2 + 4D^2\eta^2}.$$ (11.54)

11.4 Vibration Isolation

Vibratory forces generated by machines and engines are often unavoidable; however, they can be reduced substantially by properly designed springs, which we refer to as isolators.

Let the force Q acting on the system of Fig. 11.3 be the excitation source to be isolated by the spring c. The transmitted force through the spring and the damper is

$$Q_T = c\hat{q}\sqrt{1 + 4D^2\eta^2}\,. \tag{11.55}$$

Since the amplitude $\hat{q}$ developed under the force $Q(t) = \hat{Q}\cos\Omega t$ is given by Eq. (11.43)$_1$, the above equation reduces to

$$Q_T = \hat{Q}\sqrt{1 + 4D^2\eta^2}\,V_a(\eta)\,. \tag{11.56}$$

This result indicates that the ratio of transmitted force to the exciting force is identical with the ratio of the transmitted displacement to the excited displacement of Eq. (11.52). Each of these ratios is referred to as transmissibility. This transmissibility is less than unity only for $\eta > \sqrt{2}$. The results also indicate that an undamped spring is superior to a damped spring; however, some damping may be desirable when it is necessary for Ω to pass through the point of resonance.

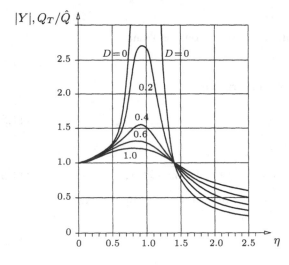

Fig. 11.7
Force transmissibility

11.5 Exercises to Chapter 11

Problem 11.1:

The system shown in the figure consists of a rigid beam, a spring, a pulley, a massless string and two point masses. The point mass m at A is not attached to the beam. In the position shown, the spring is relaxed. m, r, l, and c are given. At time $t = 0$, $\varphi = 0$ and $\dot{\varphi} = \dot{\varphi}_0$.

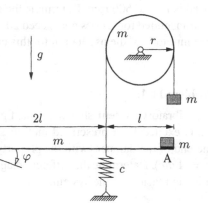

1. Derive the differential equation for small oscillations of the system. From this equation, obtain the angle $\varphi(t)$ describing the oscillation of the system in the form $\varphi(t) = \varphi_1 \cos \omega t + \varphi_2 \sin \omega t + \varphi_p$, where φ_1, φ_2 and φ_p are constants and ω is the natural circular frequency of the system.
2. Bring the equation describing the motion of the system in the form $\varphi_t = \varphi_0 \cos (\omega t + \varphi_{\text{phas}})$, where φ_0, φ_{phas} are constants.
3. From this form of the equation of motion, determine the maximum value $\dot{\varphi}_{\text{max}}$ for the angular velocity $\dot{\varphi}$ at time $t = 0$ such that the mass m at point A remains in contact with the beam during the entire course of the oscillation.

Problem 11.2:

The system shown in the figure can be regarded as the model of a motor vehicle on a sinusoidal road. For

$$l = 4 \text{ m}, \quad \frac{c}{m} = \omega_0^2 = 100 \text{ s}^{-2},$$

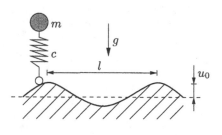

1. determine the ratio amplitude of the vertical vibration of the vehicle to amplitude of the road u_0 (x_{max}/u_0) as a function of the horizontal velocity v of the vehicle.
2. To limit this ratio, a dashpot or viscous damper with a viscous damping coefficient d is intentionally mounted parallel to the spring. Determine the damping factor D such that the ratio x_{max}/u_0 does not exceed 1.5. Here, ω_0 denotes the natural circular frequency of the system.

Problem 11.3:

A rotor of an engine with total mass $m = 1,000$ kg has the following properties: mass $m_1 = 300$ kg, and eccentricity of the center of mass $e = 0.1$ mm. The rated speed is $n = 1,500$ rpm. Determine the stiffness of the elastic foundation such that the transmitted force does not exceed 20 % of the exciting force. Moreover, compute the amplitude of the oscillation for this case.

Problem 11.4:

The vibratory system shown in the figure is excited by a prescribed motion at one end of the damper (with the property $d^2 = 4\,mc$). Determine the different magnification factors of the system.

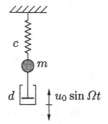

12. Systems of Several Degrees of Freedom

12.1 A Typical Example

There are two methods available for obtaining the equations of motion of a system. The first is by the application of Newton's second law, or alternatively by using d'Alembert's principle, considering forces, masses, and accelerations. The second utilizes Lagrange's equations. A typical example may be described as follows:

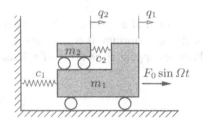

Fig. 12.1
A typical two degrees of freedom system

Figure 12.1 shows a large body m_1 able to move horizontally, a motion described by q_1 but restrained by a spring of stiffness c_1. Connected to m_1 is another body m_2, which moves with relative motion q_2, so that the force in the connecting spring is $c_2 q_2$. To the body m_1 an external independent force $F_o \sin \Omega t$ is applied horizontally. Then from Newton's law the following system of differential equations

$$m_1 \ddot{q}_1 + c_1 q_1 - c_2 q_2 = F_0 \sin \Omega t$$
$$m_2 (\ddot{q}_1 + \ddot{q}_2) + c_2 q_2 = 0$$

(12.1)

describes the motion of the two bodies.

Applying now Lagrange's equations, we introduce

$$E = \frac{1}{2} m_1 \dot{q}_1^2 + \frac{1}{2} m_2 (\dot{q}_1 + \dot{q}_2)^2 ,$$

(12.2)

$$\Phi = \frac{1}{2} c_1 q_1^2 + \frac{1}{2} c_2 q_2^2 - F_0 \sin \Omega t\, q_1 .$$

(12.3)

So, from Eq. (10.19) for a conservative system, we find

$$\frac{\partial E}{\partial \dot{q}_1} = m_1 \dot{q}_1 + m_2(\dot{q}_1 + \dot{q}_2),$$

$$\frac{\partial E}{\partial \dot{q}_2} = m_2(\dot{q}_1 + \dot{q}_2),$$

$$\frac{\partial E}{\partial q_i} = 0, \quad i = 1, 2$$

(12.4)

$$\frac{\partial \Phi}{\partial q_1} = c_1 q_1 - F_0 \sin \Omega t, \quad \frac{\partial \Phi}{\partial q_2} = c_2 q_2,$$

and thus

$$m_1 \ddot{q}_1 + m_2(\ddot{q}_1 + \ddot{q}_2) + c_1 q_1 = F_0 \sin \Omega t$$
$$m_2(\ddot{q}_1 + \ddot{q}_2) \qquad\qquad + c_2 q_2 = 0.$$

(12.5)

It is easily seen that Eqs. (12.5) and (12.1) are linear combinations of each other.

12.2 General Equations and Solution

To find the frequencies and modes of small oscillation of a system with n degrees of freedom (DOF) about a configuration of stable equilibrium, it is not necessary to consider any damping. The equations will always appear as an array of n linear equations of second order

$$\boxed{\mathbf{M}\ddot{\mathbf{q}} + \mathbf{C}\mathbf{q} = \mathbf{0}}$$

(12.6)

for free vibration, and

$$\boxed{\mathbf{M}\ddot{\mathbf{q}} + \mathbf{C}\mathbf{q} = \mathbf{F}(t)}$$

(12.7)

for forced vibration, where $\mathbf{M} = \mathbf{M}^T$ is a symmetric mass matrix. For conservative systems it turns out that the stiffness matrix $\mathbf{C} = \mathbf{C}^T$ is symmetric too. The coefficients m_{ij} ($i \neq j$, and if they exist) are known as dynamic coupling coefficients; and the coefficients c_{ij} ($i \neq j$) are termed coefficients of static coupling.

Solutions to the above set of equations (12.6) for free vibration are of the form

$$\mathbf{q}(t) = \hat{\mathbf{q}} \cos(\omega_0 t + \varphi),$$

(12.8)

in which the eigenfrequency ω_0 and the phase angle φ are the same for all q_i. Substitution verifies this and produces a system of homogeneous equations

$$\left\{ -\omega_0^2 \mathbf{M} + \mathbf{C} \right\} \hat{\mathbf{q}} \cos(\omega_0 t + \varphi) = 0.$$

(12.9)

If the amplitudes $\hat{q}_i$ are not all zero, the coefficient determinant of these equations must vanish

$$\boxed{\det\left\{ -\omega_0^2 \mathbf{M} + \mathbf{C} \right\} = 0.}$$

(12.10)

This determinant, which is called the Lagrangian determinant, may then be expanded, and yields a polynomial of nth degree in ω_0^2, called the frequency equation.

The n roots are the n values of ω_0, the n eigenvalues of the system. They are usually arranged in order of ascending magnitude, the lowest being called the first natural frequency, etc.

Equation (12.10) has been derived from a system of n equations: there remain $n - 1$ independent equations which establish the relative magnitudes of the amplitudes $\hat{q}_i$. For example, suppose $\hat{q}_1$ is chosen as the one arbitrary constant: the $n - 1$ ratios $\hat{q}_2/\hat{q}_1, \ldots, \hat{q}_i/\hat{q}_1$ may be determined from the equations (12.9) after substitution of each value of ω_0^2. For each natural frequency the different values of the $\hat{q}_i/\hat{q}_1$ corresponding to that frequency describe the mode of vibration. Thus with the lowest eigenfrequency, the system oscillates in its first mode; and so forth.

12.3 Forced Vibration with Harmonic Excitation

Let us consider the forced vibration of a typical undamped two DOF system such as shown in Fig. 12.1. The equations of motion of this system are (see Eqs. 12.1)

$$m_1 \ddot{q}_1 + c_1 q_1 - c_2 q_2 = F_0 \sin \Omega t$$
$$m_2 (\ddot{q}_1 + \ddot{q}_2) + c_2 q_2 = 0.$$
(12.11)

As with a system having a single DOF, the steady-state forced vibration will have a circular frequency agreeing with the excitation

$$q_1 = \hat{q}_1 \sin \Omega t, \quad q_2 = \hat{q}_2 \sin \Omega t.$$
(12.12)

Introducing these into the above equations, we find

$$(-m_1 \Omega^2 + c_1) \hat{q}_1 \qquad -c_2 \hat{q}_2 = F_0$$
$$-m_2 \Omega^2 \hat{q}_1 + (-m_2 \Omega^2 + c_2) \hat{q}_2 = 0,$$
(12.13)

two equations for the two unknown amplitudes $\hat{q}_1, \hat{q}_2$; with the solutions

$$\hat{q}_1 = F_0 (c_2 - m_2 \Omega^2)/D$$
$$\hat{q}_2 = F_0 m_2 \Omega^2/D,$$
(12.14)

where D is the Lagrangian determinant of the free system, that is

$$D = \begin{vmatrix} c_1 - m_1 \Omega^2 & -c_2 \\ -m_2 \Omega^2 & c_2 - m_2 \Omega^2 \end{vmatrix}.$$
(12.15)

But it is known that this determinant vanishes when Ω equals either ω_{01} or ω_{02}, the two natural frequencies; and, at these two frequencies, both $\hat{q}_1$ and $\hat{q}_2$ become infinite, i.e. there are two resonances (see Fig. 12.2).

However, from Eq. (12.14), we also realize that the amplitude $\hat{q}_1$ may vanish for

$$c_2 = m_2 \Omega^2.$$
(12.16)

Thus the response of the principal mass m_1 may be reduced to zero, so that is does not move at all. This describes the behaviour of a dynamic absorber for the forced

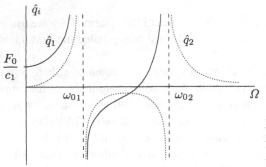

Fig. 12.2
Amplitudes $\hat{q}_1$, $\hat{q}_2$ vs excitation
frequency Ω; with the natural
frequencies ω_{01}, ω_{02}.

vibration of mass m_1 under harmonic excitation, and with a frequency Ω close to
the resonance of the system c_1, m_1. Adding a second system c_2, m_2 to this system
close to resonance allows to determine c_2 and m_2 such that the amplitude $\hat{q}_1$ may
vanish.

Example 12.1:

The horizontal oscillation of a tower may
be described by the oscillation of a spring-
mass-system of one DOF. Due to some harmonic excitation with the frequency Ω, this
system undergoes a forced vibration close
to the resonance of the system. To improve
the situation, a pendulum as dynamic absorber has to be added to the system. Determine the dimensions of the pendulum.

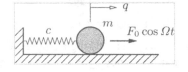

Solution:

From Eq. (10.19), describing the motion of the system are the Lagrange equations

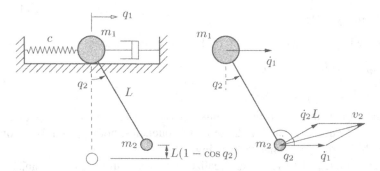

Fig. 12.3 System with velocities

$$\frac{\mathrm{D}}{\mathrm{d}t}\left(\frac{\partial E}{\partial \dot{q}_i}\right) - \frac{\partial E}{\partial q_i} = Q_i, \quad (i = 1, 2).$$

The kinetic energy of the system is

$$
E = \frac{1}{2} m_1 \dot{q}_1^2 + \frac{1}{2} m_2 v_2^2 = \sum_{i=1}^{2} \frac{1}{2} m_i v_i^2
$$
$$
= \frac{1}{2} m_1 \dot{q}_1^2 + \frac{1}{2} m_2 \left[(\dot{q}_1 + L\dot{q}_2 \cos q_2)^2 + (L\dot{q}_2 \sin q_2)^2 \right]
$$
$$
= \frac{1}{2} m_1 \dot{q}_1^2 + \frac{1}{2} m_2 \left[\dot{q}_1^2 + L^2 \dot{q}_2^2 + 2L\dot{q}_1 \dot{q}_2 \cos q_2 \right].
$$

Thus we find (for small values of q_2, i.e. $q_2 \ll 1$)

$$
\boxed{E = \frac{1}{2} (m_1 + m_2) \dot{q}_1^2 + \frac{1}{2} m_2 L^2 \dot{q}_2^2 + m_2 L\dot{q}_1 \dot{q}_2 .}
$$

The generalized forces Q_i are determined from the virtual work of the forces acting in this system under an admissible virtual displacement, since

$$
\delta A = \sum_i Q_i \, \delta q_i
$$

defines these generalized forces (see Eq. 10.10)

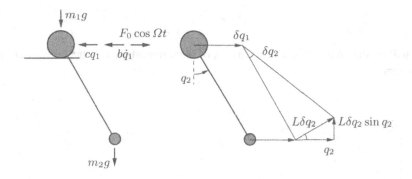

Fig. 12.4 System with forces (left) and virtual displacements (right)

$$
\delta A = (F_0 \cos \Omega t - cq_1 - b\dot{q}_1) \, \delta q_1 - m_2 g L q_2 \, \delta q_2 = Q_1 \, \delta q_1 + Q_2 \, \delta q_2 .
$$

We may calculate the following vectorial quantities

$$
\frac{\partial E}{\partial \dot{q}_i} = \left\{ \begin{array}{c} (m_1 + m_2) \dot{q}_1 + m_2 L\dot{q}_2 \\ m_2 L\dot{q}_1 + m_2 L^2 \dot{q}_2 \end{array} \right\}, \qquad \frac{\partial E}{\partial q_i} \equiv 0 .
$$

Introducing this into the Lagrange equations, we find

$$
\begin{pmatrix} (m_1 + m_2) & m_2 L \\ m_2 L & m_2 L^2 \end{pmatrix} \ddot{\mathbf{q}} + \begin{pmatrix} b & 0 \\ 0 & 0 \end{pmatrix} \dot{\mathbf{q}} + \begin{pmatrix} c & 0 \\ 0 & m_2 g L \end{pmatrix} \mathbf{q} = \begin{pmatrix} F_0 \cos \Omega t \\ 0 \end{pmatrix}.
$$

We first determine the eigenfrequencies and the modes of vibration for an undamped system.

From Eq. (12.10), we conclude

$$
\begin{vmatrix}
c - \omega_0^2(m_1 + m_2) & -\omega_0^2 m_2 L \\
-\omega_0^2 m_2 L & m_2 g L - \omega_0^2 m_2 L^2
\end{vmatrix} = 0
$$

with the frequency equation

$$
\omega_0^4 - \omega_0^2 \left[\bar{\omega}_1^2 + \bar{\omega}_2^2 \left(1 + \frac{m_2}{m_1}\right)\right] + \bar{\omega}_1^2 \bar{\omega}_2^2 = 0 ,
$$

where $\bar{\omega}_1^2 = c/m_1$ and $\bar{\omega}_2^2 = g/L$, respectively, are the squares of the eigenfrequencies of the two single DOF systems: initial spring-mass-system and pendulum.

The solutions of the frequency equation are ($\xi = m_2/m_1$)

$$
\omega_{01,2}^2 = \frac{1}{2}\left[\bar{\omega}_1^2 + (1 + \xi)\bar{\omega}_2^2\right] \pm \frac{1}{2}\sqrt{\bar{\omega}_1^4 + (1 + \xi)^2\,\bar{\omega}_2^4 - 2(1 - \xi)\,\bar{\omega}_1^2 \bar{\omega}_2^2} ,
$$

and moreover

$$
\boxed{\omega_{01,2}^2 = \frac{1}{2}\bar{\omega}^2 \left[2 + \xi \pm \sqrt{4\xi + \xi^2}\,\right]}
$$

for the special case of identical eigenfrequencies of the underlying initial systems, i.e.

$$
\bar{\omega}_1^2 = \bar{\omega}_2^2 = \bar{\omega}^2 .
$$

It will be shown that this special case is of great importance for the behaviour of our system.

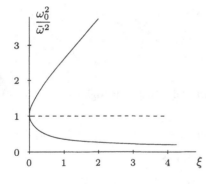

Fig. 12.5
Solution of eigenfrequencies vs mass relation ξ

We now design the dynamic absorber. For the forced vibration of the (undamped) system, we find

$$
\begin{pmatrix}
c - \Omega^2(m_1 + m_2) & -\Omega^2 m_2 L \\
-\Omega^2 m_2 L & m_2 g L - \Omega^2 m_2 L^2
\end{pmatrix} \hat{\mathbf{q}} = \begin{pmatrix} Q_0 \\ 0 \end{pmatrix} .
$$

From the second equation, we determine

$$
-\Omega^2 \hat{q}_1 + (g - \Omega^2 L)\,\hat{q}_2 = 0 \quad \rightarrow \quad \hat{q}_1 = \frac{g - \Omega^2 L}{\Omega^2}\,\hat{q}_2 ,
$$

and thus from the first equation

$$\left[c - \Omega^2(m_1 + m_2)\right] \frac{g/L - \Omega^2}{\Omega^2} \hat{q}_2 L - \Omega^2 m_2 L \hat{q}_2 = Q_0,$$

and finally

$$\hat{q}_2 L = \frac{Q_0}{c} \frac{\eta_2^2}{(1 - \eta_2^2)\left[1 - \eta_1^2(1 + \xi)\right] - \eta_1^2 \eta_2^2 \xi} = \frac{Q_0}{c} V_2(\eta_1, \eta_2),$$

where

$$\eta_1 = \frac{\Omega}{\sqrt{c/m_1}} \cong 1, \quad \eta_2 = \frac{\Omega}{\sqrt{g/L}},$$

as well as

$$\hat{q}_1 = \frac{Q_0}{c} \frac{1 - \eta_2^2}{\eta_2^2} V_2(\eta_1, \eta_2) = \frac{Q_0}{c} V_1(\eta_1, \eta_2).$$

It turns out that there exists a solution to our problem with $\hat{q}_1 = 0$, namely for

$$\eta_2 = \frac{\Omega}{\sqrt{g/L}} = 1 \quad \rightarrow \quad \frac{g}{L} = \Omega^2 \quad \rightarrow \quad \boxed{L \cong g \frac{m_1}{c}.}$$

The amplitude of the motion of the pendulum is then given by

$$\hat{q}_2 L = \frac{Q_0}{c} V_2(\eta_1, \eta_2 = 1) = -\frac{Q_0}{c} \frac{1}{\xi \eta_1^2} = -Q_0 \frac{L}{m_2 g} = -Q_0 \frac{1}{m_2 \Omega^2}.$$

To get a small amplitude, we choose a large mass m_2.

12.4 Exercises to Chapter 12

Problem 12.1:
Compute the natural frequencies for the massless frame shown in the figure, with a concentrated mass at its free end, and with constant flexural rigidity EJ.

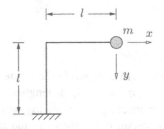

Problem 12.2:
Two identical oscillators (each with mass m and length l) are connected by an elastic spring as shown in the figure. Compute the natural frequencies for small oscillations.

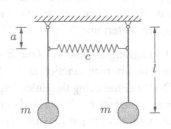

Problem 12.3:

For the system shown in the figure with constant flexural rigidity EJ

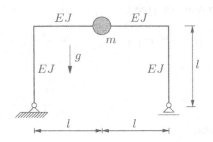

1. derive using the influence coefficients δ_{ik} the differential equation for small oscillations of the system.
2. Establish the eigenvalue problem,
3. and determine the natural frequencies of the system.

Problem 12.4:

Solve Problem 12.3 using the Euler-Lagrange equations for conservative systems.

Problem 12.5:

As shown in the figure, a rigid disk of diameter d, mass m, and mass moment of inertia θ, is fixed to a massless flexible shaft of constant flexural rigidity. Compute the natural frequencies, and the different modes of vibration.

$l = d = 1\,\mathrm{m}, \quad G = 7.85\,\mathrm{kN}, \quad EJ = 400\,\mathrm{kNm^2}$

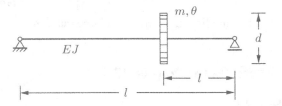

Problem 12.6:

For the system depicted in the figure, the masses m_1, m_2 and m_3, the length l of the strings connecting m_1 with m_2 and the spring constant c are given. In the position $q_3 = 0$, the spring is relaxed. Damping and friction forces are assumed to be negligible. Determine

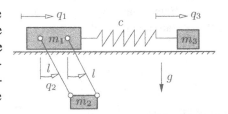

1. the Lagrangian function $L = E - \Pi$
2. the Hamiltonian function $H = E + \Pi$
3. Derive either using the Euler-Lagrange equations or directly from the Hamiltonian function the differential equations describing the motion of the system if the motion of m_1 is given by $q_1 = \frac{1}{2}at^2$.

13. Answers to the Exercises

Chapter 1: Basic Concepts of Continuum Mechanics

Problem 1.2:

The eigenvalues are

σ_a: $\sigma_1 = 30$, $\sigma_2 = -6$, $\sigma_3 = 12$

σ_b: $\sigma_1 = 25$, $\sigma_2 = 20$, $\sigma_3 = 10$

σ_c: $\sigma_1 = 40$, $\sigma_2 = 40$, $\sigma_3 = 0$

ε_a: $\varepsilon_1 = -10^{-4}$, $\varepsilon_2 = 10^{-4}$, $\varepsilon_3 = 4 \cdot 10^{-4}$

ε_b: $\varepsilon_1 = 3 \cdot 10^{-4}$, $\varepsilon_2 = 2 \cdot 10^{-4}$, $\varepsilon_3 = -10^{-4}$

ε_c: $\varepsilon_1 = 10^{-4}$, $\varepsilon_2 = 10^{-4}$, $\varepsilon_3 = -2 \cdot 10^{-4}$

Problem 1.3:

a) $\sigma_1 = 178.06$, $\sigma_2 = -78.06$, $\varphi_0 = 19.33°$

b) $\sigma_{\bar{x}\bar{x}} = 130$, $\sigma_{\bar{y}\bar{y}} = -30$, $\sigma_{\bar{x}\bar{y}} = -100$

c) $\tau_{\max} = 128.06$, $\varphi_1 = 64.33°$, $\sigma_{\bar{x}\bar{x}} = \sigma_{\bar{y}\bar{y}} = 50$,

Problem 1.4:

1. $\varepsilon = \dfrac{1}{l_0} \begin{pmatrix} 1 & 3 \\ 3 & -3 \end{pmatrix}$, 2. $\alpha = \dfrac{1}{l_0} \begin{pmatrix} 0 & 0 \\ 0 & 0 \end{pmatrix}$

3. $\varphi_0 = 28.15°$, $\varepsilon_1 l_0 = -1 - \sqrt{13}$, $\varepsilon_2 l_0 = -1 + \sqrt{13}$

Problem 1.5:

$$\varepsilon_{xx} = \varepsilon_a, \quad \varepsilon_{yy} = \frac{1}{3}(2\varepsilon_b + 2\varepsilon_c - \varepsilon_a), \quad \varepsilon_{xy} = \frac{1}{\sqrt{3}}(\varepsilon_b - \varepsilon_c)$$

$$E_1 = \frac{2}{3}(\varepsilon_a + \varepsilon_b + \varepsilon_c), \quad E_2^* = \frac{1}{3}(\varepsilon_a^2 + \varepsilon_b^2 + \varepsilon_c^2 - 2\varepsilon_a\varepsilon_b - 2\varepsilon_b\varepsilon_c - 2\varepsilon_c\varepsilon_a)$$

$$\varepsilon_{1,2} = \frac{1}{3}(\varepsilon_a + \varepsilon_b + \varepsilon_c) \pm \frac{2}{3}\sqrt{\varepsilon_a^2 + \varepsilon_b^2 + \varepsilon_c^2 - \varepsilon_a\varepsilon_b - \varepsilon_b\varepsilon_c - \varepsilon_c\varepsilon_a}$$

Chapter 2: Elastic Material

Problem 2.1:

1. $\sigma_{eq} = \dfrac{1}{2}\sigma + \dfrac{1}{2}\sqrt{\sigma^2 + 4\tau^2}$

2. $\sigma_{eq} = \sqrt{\sigma^2 + 4\tau^2}$

3. $\sigma_{eq} = \sqrt{\sigma^2 + 2\left(1+\nu\right)\tau^2}$

4. $\sigma_{eq} = \sqrt{\left(\sigma^2 + 3\,\tau^2\right)}$

Problem 2.2:

$$W = \frac{F^2 a}{EA}\left(6 + 5\sqrt{2}\right)$$

Problem 2.3:

$$\delta_H = -\frac{Fa(4 + 2\sqrt{2})}{EA}, \quad \delta_V = \frac{Fa(8 + 6\sqrt{2})}{EA}$$

Problem 2.4:

$$\delta_H = 4\alpha\theta_0 a, \quad \delta_V = 0$$

Problem 2.5:

$$\delta_V = -\alpha\Theta a, \quad \delta_H = 0$$

Problem 2.6:

$$\delta_V = \frac{2(2 + \sqrt{2})Fa}{EA}, \quad \delta_H = -\frac{2Fa}{EA}$$

Problem 2.7:

$$\delta_H = \frac{Fa}{2EA}\left(\sqrt{2} - 1\right), \quad \delta_V = \frac{Fa}{2EA}\left(3 - \sqrt{2}\right)$$

Chapter 3: The Theory of Simple Beams I

Problem 3.1:

a) $\varphi_0 = 0°, \quad J_1 = J_{zz} = \dfrac{1}{3}a^3\delta, \quad J_2 = J_{yy} = \dfrac{1}{12}a^3\delta$

b) $\varphi_0 = 45°, \quad J_1 = \dfrac{7}{6}a^3\delta, \quad J_2 = \dfrac{1}{6}a^3\delta$

Problem 3.2:

$a = 16.68$ cm

Problem 3.3:

$$\sigma_{eq} = \frac{20F}{b^2} \left\{ 1 - 12\frac{z}{b} + \sqrt{\left(1 - 12\frac{z}{b}\right)^2 + \frac{9}{100}\left(1 - 4\frac{z^2}{b^2}\right)^2} \right\}$$

2. $\sigma_{eq}(0,0,0) = (20 + 2\sqrt{109})\dfrac{F}{b^2}$

3. $\sigma_{eq}(0,0,\dfrac{b}{4}) = (-40 + \dfrac{1}{2}\sqrt{6481})\dfrac{F}{b^2}$

4. $\sigma_{eq}(0,0,\dfrac{b}{2}) = 240\dfrac{F}{b^2}$

Problem 3.4:

$\varphi = 26.45°$

$\sigma_{max,min} = \pm 51.41$ N/mm^2

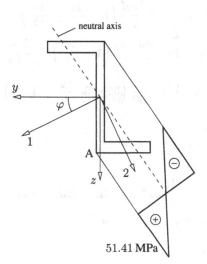

Problem 3.5:

$$\sigma_{max} = \frac{55}{28}\frac{F}{\delta a}, \quad \sigma_{min} = -\frac{41}{28}\frac{F}{\delta a}$$

Problem 3.6:

a) $\tau_{max} = \dfrac{2Q}{3h_m\delta}, \quad h_m = a\dfrac{\sqrt{3}}{2}, \quad \tau_1 = \dfrac{3}{4}\tau_{max}$

b) $\tau_{max} = \dfrac{3Q}{2\sqrt{2}a\delta}$

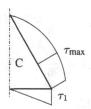

Problem 3.7:

a) $\tau_{max} = \dfrac{3Q}{2h\delta_1}\dfrac{h\delta_1 + 4b\delta_2}{h\delta_1 + 6b\delta_2}$, $e = \dfrac{3b^2\delta_2}{h\delta_1 + 6b\delta_2}$

b) $\tau_{max} = \dfrac{3Q}{2\sqrt{2}a\delta}$, $e = 0$

c) $\tau_{max} = \dfrac{3Q}{\sqrt{2}a\delta}$, $e = \dfrac{a}{2\sqrt{2}}$

d) $\tau_{max} = \dfrac{2Q\,(a+r)}{r\,\delta\,(4\,a+\pi\,r)}$, $e = \dfrac{2(2r^2 + a^2 + \pi ar)}{4a + \pi r}$

Problem 3.8:

a) $EJw_1(x_1) = \dfrac{Mx_1^2}{4}\left(1 - \dfrac{x_1}{2l}\right)$,

$EJw_2(x_2) = \dfrac{Mx_2 l}{8}\left(1 - \dfrac{x_2}{l} - \dfrac{x_2^2}{l^2}\right)$

b) $EJw_1(x_1) = \dfrac{q_0 l^4}{72}\dfrac{x_1}{l}\left(4 - \dfrac{x_1^2}{l^2}\right) - \dfrac{M_0 l^2}{12}\dfrac{x_1}{l}\left(8 - 6\dfrac{x_1}{l} + \dfrac{x_1^2}{l^2}\right)$,

$EJw_2(x_2) = \dfrac{q_0 l^4}{360}\left(52 - 55\dfrac{x_2}{l} + 3\dfrac{x_2^5}{l^5}\right) - \dfrac{M_0 l^2}{3}\left(1 - \dfrac{x_2}{l}\right)$

Problem 3.9:

$c_\theta = \dfrac{3EJA\,\alpha\theta_0}{3J + Al^2}$

$EJw_1(x_1) = \dfrac{1}{6}\,c_\theta x_1^2(x_1 - 3l), \quad EJw_2(x_2) = -\dfrac{1}{6}\,c_\theta(x_2 - l)^2(x_2 + 2l)$

$M_s = c_\theta l, \quad Q_s = N_s = -c_\theta$

Problem 3.10:

$w = w_q + w_\theta$,

$w_q = \dfrac{qx^2}{24EJ}\,(6l^2 - 4lx + x^2), \quad w_\theta = -\alpha\dfrac{T_2 - T_1}{2h}\,x^2$

Problem 3.11:

$EJw_1(x_1) = \dfrac{5}{96}\,Fl^2 x_1\left(5 - 2\dfrac{x_1^2}{l^2}\right), \quad l = 5a$

$EJw_2(x_2) = \dfrac{5}{96}\,Fl^3\left(5 - \dfrac{x_2}{l} - 6\dfrac{x_2^2}{l^2} + 2\dfrac{x_2^3}{l^3}\right)$

Problem 3.12:

$w(x) = \dfrac{1}{2k}\,(x - l)^2$

Chapter 4: Torsion of Prismatic Bars

Problem 4.1:

a) $|\tau|_{max} = \dfrac{M}{2\pi r^2 \delta}$

b) $|\tau|_{max} = \dfrac{2M}{\sqrt{3}a^2 \delta}$

c) $|\tau|_{max} = \dfrac{M}{2b^2 \delta}$

With a constant stress across the thickness of the profile.

Problem 4.2:

a) $|\tau|_{max} = \dfrac{3M}{2\pi r \delta^2}$

b) $|\tau|_{max} = \dfrac{M}{\beta a \delta^2}$

c) $|\tau|_{max} = \dfrac{M}{3\beta b \delta^2}$

With a linearly varying stress across the thickness of the profile.

Problem 4.3:

As sum of the shear stresses due to the shear force and the additional torque, we arrive at:

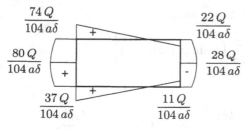

$\dfrac{74\,Q}{104\,a\delta}$ $\dfrac{22\,Q}{104\,a\delta}$

$\dfrac{80\,Q}{104\,a\delta}$ $\dfrac{28\,Q}{104\,a\delta}$

$\dfrac{37\,Q}{104\,a\delta}$ $\dfrac{11\,Q}{104\,a\delta}$

Problem 4.4:

The distributions of the shear stresses are:

1. due to the shear force Q

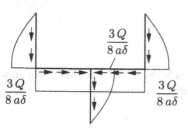

$\dfrac{3\,Q}{8\,a\delta}$

$\dfrac{3\,Q}{8\,a\delta}$ $\dfrac{3\,Q}{8\,a\delta}$

2. total shear stress at b-b

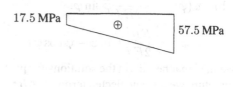

17.5 MPa   57.5 MPa

Problem 4.5:

$$\frac{\tau_o}{\tau_\square} = \frac{\pi}{4}, \quad \frac{\varphi_o}{\varphi_\square} = \frac{\pi^2}{16}$$

Problem 4.6:

a) $|\tau|_{max} = 18.86\ \text{N/mm}^2,\ \varphi_A = 6.21 \cdot 10^{-3}$

b) $|\tau|_{max} = 17.64\ \text{N/mm}^2,\ \varphi_A = 4.35 \cdot 10^{-3}$

c) $|\tau|_{max} = 74.07\ \text{N/mm}^2,\ \varphi_A = 28.16 \cdot 10^{-3}$

d) $|\tau|_{max} = 60.06\ \text{N/mm}^2,\ \varphi_A = 21.21 \cdot 10^{-3}$

Chapter 5: Curved Beams

Problem 5.1:

a) $w(\varphi) = c_1 \cos\varphi + c_2 \sin\varphi - \dfrac{FR^3}{2EJ}\varphi\cos\varphi, \quad c_1 = 0, \quad c_2 = \dfrac{FR^3}{2EJ},$

$$u(\varphi) = \frac{FR^3}{2EJ}\left(\varphi\sin\varphi - 2 + 2\cos\varphi\right)$$

b) $w(\varphi) = \dfrac{FR^3}{2EJ}\left[\left(\dfrac{\pi}{2} - 1 - \dfrac{1}{\pi}\right)\sin\varphi + \cos\varphi\right.$

$$\left. -1 + \frac{\varphi}{2\pi}\left(\pi\sin\varphi - 2\cos\varphi\right)\right],$$

$$u(\varphi) = \frac{FR^3}{2EJ}\left[\left(\frac{\pi}{2} - 1\right)\cos\varphi - \frac{3}{2}\sin\varphi\right.$$

$$\left. +1 - \frac{\pi}{2} + \frac{\varphi}{2\pi}\left(\pi\sin\varphi + 2\cos\varphi\right)\right]$$

Problem 5.2:

a) $w(\varphi) = c_1 \sin\varphi + c_2 \cos\varphi + \dfrac{MR^2}{EJ}, \quad c_1 = 0, \quad c_2 = -\dfrac{MR^2}{EJ},$

$$u(\varphi) = -\frac{MR^2}{EJ}\left(\varphi - \sin\varphi\right)$$

b) $w(\varphi) = -\dfrac{FR^3}{2EJ}\varphi\sin\varphi,$

$$u(\varphi) = \frac{FR^3}{2EJ}\left(\sin\varphi - \varphi\cos\varphi\right)$$

We note that herein (in the solutions for problems 5.1b and 5.2a) for beams of small curvature we have neglected terms J/AR^2 compared with one.

Chapter 6: Simple Beams II: Energy Principles

Problem 6.1:

$$W^* = (3 + \sqrt{2})\frac{F^2 a}{AE}, \quad \delta_V = \frac{Fa}{AE}, \quad \delta_H = (5 + 2\sqrt{2})\frac{Fa}{AE}$$

Problem 6.2:

$$\delta_V = 11\frac{Fa}{EA}$$

Problem 6.3:

$$\delta_V = \frac{137ql^4}{3EJ}, \quad \delta_H = -\frac{18ql^4}{EJ}$$

Problem 6.4:

The distribution of the bending moments is

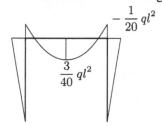

Problem 6.5:

The distribution of the bending moments is

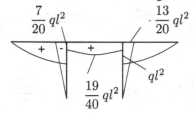

Problem 6.6:

$$S_1 = \frac{6F}{7}, \quad S_2 = \frac{F}{7}, \quad S_3 = \frac{2F}{7}, \quad S_4 = \frac{-2F}{7}, \quad S_5 = \frac{-F}{7}$$

$$S_6 = \frac{-6F}{7}, \quad S_7 = \frac{-F}{7}, \quad S_8 = \frac{F}{7}, \quad S_9 = F, \quad S_{10} = 0$$

$$S_{11} = 0, \quad S_{12} = -F, \quad S_{13} = F, \quad S_{14} = -F, \quad S_{15} = 0$$

Problem 6.7:

$$N = -\frac{3EJ\alpha\Theta}{l^2 + 3\frac{EJ}{EA}}$$

Problem 6.8:

We introduce the two redundants X_1, X_2 and compute

$$X_1 = \frac{4F}{223}(15 + \sqrt{2}), \quad X_2 = -\frac{F}{223}(24 - 43\sqrt{2})$$

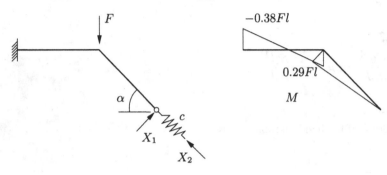

Problem 6.9:

$$\varphi = \left(\frac{1}{2} - \frac{\pi}{4}\right)\frac{FR^2}{EJ}$$

Problem 6.10:

1. $\varphi = \left(\pi + \frac{1}{2}\right)\frac{FR^2}{EJ}$

2. $\sigma_{xx} = \frac{F}{a^2}, \quad \sigma_{xz} = \frac{3F}{2a^2}, \quad \sigma_{1,2} = \frac{1}{2}\frac{F}{a^2}(1 \pm \sqrt{10}), \quad \varphi = -35.78°$

Problem 6.11:

$$S_1 = -\frac{2F_1}{3}, \quad S_2 = -\frac{2F_1}{3}, \quad S_3 = \frac{2\sqrt{2}F_1}{3}, \quad S_4 = 0, \quad S_5 = -\frac{5F_1}{6}$$

$$S_6 = -\frac{F_1}{2}, \quad S_7 = -\frac{\sqrt{2}F_1}{6}, \quad S_8 = \frac{\sqrt{2}F_1}{6}, \quad S_9 = \frac{F_1}{2}, \quad S_{10} = \frac{-F_1}{6}$$

$$S_{11} = -\frac{F_1}{3}, \quad S_{12} = \frac{\sqrt{2}F_1}{3}, \quad S_{13} = 0, \quad S_{14} = -\frac{F_1}{3},$$

Chapter 7: Two-dimensional Problems

Problem 7.1:

a) $F = \dfrac{p_x}{2h} y^2$

b) $F = \dfrac{p_x}{2h} \left[\dfrac{1}{2} y^2 + \dfrac{1}{3b} y^3 \right]$

Problem 7.2:

a) $F = -\dfrac{1}{2} \sigma_0 r^2$

b) $F = c_1 \ln r + c_2 r^2, \quad c_1 = \dfrac{a^2 b^2 (p_i - p_o)}{(b^2 - a^2)}, \quad 2c_2 = \dfrac{(a^2 p_i - p_o b^2)}{(a^2 - b^2)},$

$p_o = -\sigma_0, \quad p_i = 0$

Problem 7.3:

$$F(r) = a_1 \ln r + a_2 r^2 + a_3 r^2 \ln r$$

$$a_1 = \dfrac{4M}{N} \left(a^2 b^2 \ln(b/a) \right)$$

$$a_2 = -\dfrac{M}{N} \left[b^2 - a^2 + 2(b^2 \ln b - a^2 \ln a) \right]$$

$$a_3 = \dfrac{2M}{N} (b^2 - a^2)$$

$$N = t \left[(b^2 - a^2)^2 - 4b^2 a^2 (\ln(b/a))^2 \right]$$

Problem 7.4:

$$u_x = \dfrac{1+\nu}{2E} \left(\nu x^2 + (1-\nu)z^2 \right) a, \quad M = a \int_A x^2 \, dA = aJ,$$

$$u_y = 0, \quad u_z = -\dfrac{1-\nu^2}{E} axz$$

Problem 7.5:

$$u_x = \dfrac{\tau}{G} y, \quad u_y = u_z = 0$$

Problem 7.6:

$$\sigma_{xx} = 12 \dfrac{Q}{bh^3} xy, \quad \sigma_{yy} = 0, \quad \sigma_{xy} = \dfrac{3}{2} \dfrac{Q}{bh} \left(1 - \dfrac{4y^2}{h^2} \right)$$

Problem 7.7:

$$\sigma_{rr} = -\frac{2Q}{\pi b r} \sin \varphi, \quad \sigma_{\varphi\varphi} = 0, \quad \sigma_{r\varphi} = 0$$

Problem 7.8:

$$\sigma_{rr} = \frac{\sigma_0 R^2}{2} \left[\frac{1}{R^2} - \frac{1}{r^2} + \left(\frac{1}{R^2} - \frac{4}{r^2} + 3 \frac{R^2}{r^4} \right) \cos 2\varphi \right]$$

$$\sigma_{\varphi\varphi} = \frac{\sigma_0 R^2}{2} \left[\frac{1}{R^2} + \frac{1}{r^2} - \left(\frac{1}{R^2} + 3 \frac{R^2}{r^4} \right) \cos 2\varphi \right]$$

$$\sigma_{r\varphi} = \frac{\sigma_0 R^2}{2} \left[-\frac{1}{R^2} - \frac{2}{r^2} + 3 \frac{R^2}{r^4} \right] \sin 2\varphi$$

$$\sigma_{\varphi\varphi\,\text{max}} = 3\sigma_0, \quad \varphi = \pi/2, \quad r = R$$

Problem 7.9:

$$K_1 = \frac{E_S E_D}{E_D(a + \Delta d/2)(1 - \nu_S)(b^2 - a^2) + E_S[(1 + \nu_D)ab^2 + (1 - \nu_D)a^3]}$$

1. $\Theta \geq \dfrac{\Delta d}{2\alpha a} = 104.17\,\text{K}$

2. $\sigma_{rr} = \dfrac{\Delta d E a}{2[a^2(1 - \nu) + b^2(1 + \nu)]} \left(1 - \dfrac{b^2}{r^2} \right)$

   $\sigma_{\varphi\varphi} = \dfrac{\Delta d E a}{2[a^2(1 - \nu) + b^2(1 + \nu)]} \left(1 + \dfrac{b^2}{r^2} \right), \quad \sigma_{r\varphi} = 0$

   $\sigma_{rr}(a) = -112.90\,\text{MPa}, \quad \sigma_{\varphi\varphi}(a) = 141.13\,\text{MPa}$

3. $\sigma_{rr\,S} = \sigma_{\varphi\varphi\,S} = -\dfrac{\Delta d}{2}(b^2 - a^2)K_1$

   $\sigma_{rr\,D} = \dfrac{\Delta d}{2} a^2(1 - b^2/r^2)K_1$

   $\sigma_{\varphi\varphi\,D} = \dfrac{\Delta d}{2} a^2(1 + b^2/r^2)K_1$

   $\sigma_{rr\,D}(a) = \sigma_{rr\,S}(a) = -97.46\,\text{MPa}, \quad \sigma_{\varphi\varphi\,D}(a) = 121.82\,\text{MPa}$

4. Take the solution of part c) and replace

   $$\nu_S \mapsto \frac{\nu_S}{1 - \nu_S}, \quad E_S \mapsto E_S \frac{1 + 2\nu_S}{(1 - \nu_S)^2}$$

   $\sigma_{rr\,D}(a) = \sigma_{rr\,S}(a) = -101.01\,\text{MPa}, \quad \sigma_{\varphi\varphi\,D}(a) = 126.26\,\text{MPa}$

Chapter 8: Plates and Shells

Problem 8.1:

1. $\quad w(r) = \dfrac{p_0 R^4}{64B} \left[\dfrac{5+\nu}{1+\nu} - 2 \dfrac{(3+\nu)}{(1+\nu)} \dfrac{r^2}{R^2} + \dfrac{r^4}{R^4} \right]$

2. $\quad \sigma_{rr}(r,z) = \dfrac{3}{4} \dfrac{p_0 R^2}{h^3} (3+\nu) \left(1 - \dfrac{r^2}{R^2} \right) z,$

$\quad \sigma_{\varphi\varphi}(r,z) = \dfrac{3}{4} \dfrac{p_0 R^2}{h^3} (1+3\nu) \left[\dfrac{3+\nu}{1+3\nu} - \dfrac{r^2}{R^2} \right] z$

Problem 8.2:

$$w(r) = \dfrac{FR^2}{16\pi B} \left[\dfrac{3+\nu}{1+\nu} \left(1 - \dfrac{r^2}{R^2} \right) + 2 \ln \dfrac{r}{R} \dfrac{r^2}{R^2} \right]$$

$$m_{rr}(r) = -(1+\nu) \dfrac{F}{4\pi} \ln \dfrac{r}{R},$$

$$m_{\varphi\varphi}(r) = -\dfrac{F}{4\pi} \left[(1+\nu) \ln \dfrac{r}{R} - 1 + \nu \right].$$

Problem 8.3:

a) Constant temperature field ϑ_0

$$u(r) = 0, \quad \sigma_{rr} = \sigma_{\varphi\varphi} = -\dfrac{1}{1-\nu} E\alpha\vartheta_0,$$

b) linearly distributed temperature field $\vartheta_1 z$

$$w(r) = 0, \quad m_{rr} = m_{\varphi\varphi} = -B(1+\nu)\alpha\vartheta_1,$$

$$\sigma_{rr} = \sigma_{\varphi\varphi} = -\dfrac{1}{1-\nu} E\alpha\vartheta_1 z.$$

The stresses will be superimposed to give:

$$\sigma_{rr} = \sigma_{\varphi\varphi} = -\dfrac{E\alpha}{1-\nu} (\vartheta_0 + \vartheta_1 z).$$

Problem 8.4:

$$w(r) = -\dfrac{q_0 R}{8B\Phi} R_0^2 \left\{ 4\Psi \ln \rho + (1-\rho^2)(\Phi + 2\Psi) + 2\Phi\rho^2 \ln \rho \right\}$$

$$q(r) = q_0 \dfrac{R}{r};$$

$$m_{rr}(r) = \dfrac{q_0 R}{2\Phi} \left\{ (1+\nu)^2 \ln \rho_0 + (1-\nu)(\rho_0^2 - \dfrac{\Psi}{\rho^2}) + (1+\nu)\Phi \ln \rho \right\},$$

$$m_{\varphi\varphi}(r) = \dfrac{q_0 R}{2\Phi} \left\{ (1+\nu)^2 \ln \rho_0 - (1-\nu)(1+\nu - \nu\rho_0^2 - \dfrac{\Psi}{\rho^2}) + (1+\nu)\Phi \ln \rho \right\}.$$

with

$$\rho_0 = \frac{R_0}{R}, \quad \rho = \frac{r}{R_0},$$

$$\Phi = 1 + \nu + (1-\nu)\rho_0^2, \quad \Psi = 1 - (1+\nu)\ln\rho_0,$$

Problem 8.5:

1. $w_{max}^K(x = a) = \dfrac{m_R a^2}{2B} + \dfrac{q_R a^3}{3B}$

2. $x = 0, \quad |\sigma_{xx}|_{max} = \dfrac{6}{h^2}(m_R + q_R a), \quad \sigma_{yy} = \nu\sigma_{xx}, \quad \sigma_{xy} = 0$

3. $w^K(x) = \dfrac{1}{6B}[3(m_R + q_R a)x^2 - q_R x^3]$

4. $w^B(x) = \dfrac{1}{6EJ}[3(M_R + Q_R a)x^2 - Q_R x^3],$

    $\dfrac{w^K}{w^B} = 0.889$ for $\nu = 0.3, \quad nbq_R = nQ_R, \quad nbm_R = nM_R$ and beams
    with a rectangular cross section.

Problem 8.6:

$$p_\vartheta = 0, \quad p_n = \gamma R(\cos\vartheta - \cos\vartheta_0),$$

$$n_{\vartheta\vartheta} = \gamma R^2 \left[\frac{1 - \cos^3\vartheta}{3\sin^2\vartheta} - \frac{1}{2}\cos\vartheta_0 \right]$$

$$n_{\varphi\varphi} = \gamma R^2 \left[\cos\vartheta - \frac{1}{2}\cos\vartheta_0 - \frac{1 - \cos^3\vartheta}{3\sin^2\vartheta} \right].$$

Problem 8.7:

$$h = \frac{1}{\vartheta_2 - \vartheta_1}[h_1\vartheta_2 - h_2\vartheta_1 + (h_2 - h_1)\vartheta]$$

$$n_{\vartheta\vartheta} = -\frac{\rho g R}{(\vartheta_2 - \vartheta_1)\sin^2\vartheta}[(h_1\vartheta_2 - h_2\vartheta_1)(\cos\vartheta_1 - \cos\vartheta)$$

$$+ (h_2 - h_1)(\sin\vartheta - \sin\vartheta_1 - \vartheta\cos\vartheta + \vartheta_1\cos\vartheta)]$$

$$n_{\varphi\varphi} = \frac{\rho g R}{(\vartheta_2 - \vartheta_1)\sin^2\vartheta}\{(h_1\vartheta_2 - h_2\vartheta_1)[\cos\vartheta_1 - \cos\vartheta(1 + \sin^2\vartheta)]$$

$$+ (h_2 - h_1)[\sin\vartheta - \sin\vartheta_1 - \vartheta\cos\vartheta(1 + \sin^2\vartheta) + \vartheta_1\cos\vartheta]\}$$

For $h_1 = h_2 = h, \quad \vartheta_1 = 0$, we find

$$n_{\vartheta\vartheta} = \frac{\rho g h R}{\sin^2\vartheta}[\cos\vartheta - 1]$$

$$n_{\varphi\varphi} = -\rho g h R\left[\cos\vartheta - \frac{1}{1 + \cos\vartheta}\right]$$

Chapter 9: Stability of Equilibrium

Problem 9.1:

1. $R = 10 \dfrac{EJ}{l^2}$, $\dfrac{R}{F_{\text{crit}}} = 10/\pi^2$

2. $R = 12 \dfrac{EJ}{l^2}$, $\dfrac{R}{F_{\text{crit}}} = 12/\pi^2$

Problem 9.2:

1. $R = \left(3 + \dfrac{3}{4}\dfrac{cl^3}{EJ}\right)\dfrac{EJ}{l^2}$, $\dfrac{R}{F_{\text{crit}}} = \dfrac{4}{\pi^2}\left(3 + \dfrac{3}{4}\dfrac{cl^3}{EJ}\right)$

2. $\tilde{F}_1 = \dfrac{4}{9}\left[39 + \kappa - \sqrt{1116 - 57\kappa + \kappa^2}\right]\dfrac{EJ}{l^2}$,

   $\dfrac{\tilde{F}_1}{F_{\text{crit}}} = \dfrac{4}{\pi^2}\dfrac{4}{9}\left[39 + \kappa - \sqrt{1116 - 57\kappa + \kappa^2}\right]$

Problem 9.3:

1. $R = 3(4 + \kappa)\dfrac{EJ}{l^2}$, $\kappa = \dfrac{c_m l}{EJ}$

2. $\tilde{F}_1 = \dfrac{2}{13}\left[15(18 + \kappa) - \sqrt{5}\sqrt{7664 - 564\kappa + 45\kappa^2}\right]\dfrac{EJ}{l^2}$

Problem 9.4:

1. $R[\tilde{w}] = \dfrac{\displaystyle\int_0^l EJ(\tilde{w}'')^2 \, dx}{\displaystyle\int_0^l \int_0^x (\tilde{w}')^2 \, d\xi \, dx}$

2. $R = 120 \dfrac{EJ}{l^3}$

3. $\tilde{w} = [a + (bx)/l](x^2/l^2)(1 - x/l)$, $b = -\dfrac{2}{3}a$, $R = 56 \dfrac{EJ}{l^3}$

Chapter 11: Systems of One Degree of Freedom

Problem 11.1:

1. $\ddot{\varphi} + \omega^2\varphi = K$, $\omega^2 = \dfrac{2}{9}\dfrac{c}{m}$, $K = \dfrac{5}{36}\dfrac{g}{l}$

   $\varphi(t) = \varphi_1\cos(\omega t) + \varphi_2\sin(\omega t) + \varphi_p$

   $\varphi_p = \dfrac{K}{\omega^2}$, $\varphi_1 = -\dfrac{K}{\omega^2}$, $\varphi_2 = \dfrac{\dot{\varphi}_0}{\omega}$

2. $\varphi(t) = \varphi_0\cos(\omega t + \varphi_{phas})$,

   $$\varphi_0 = \sqrt{\left(\dfrac{\dot{\varphi}_0}{\omega}\right)^2 + \left(\dfrac{K}{\omega^2}\right)^2}, \quad \varphi_{phas} = \arctan\left(\dfrac{\frac{\dot{\varphi}_0}{\omega}}{\frac{K}{\omega^2}}\right)$$

3. $\dot{\varphi}_{max}^2 = \omega^2\left\{\left(\dfrac{g}{3l\omega^2}\right)^2 - \left(\dfrac{K}{\omega^2}\right)^2\right\}$

Problem 11.2:

1. $\dfrac{x_{max}}{u_0} = \dfrac{\sqrt{1 + 4D^2\eta^2}}{\sqrt{(1 - \eta^2)^2 + 4D^2\eta^2}}$, $D = \dfrac{d}{2m\omega_0}$, $\eta = \dfrac{2\pi v}{l\omega_0}$

2. $D = 0.48$

Problem 11.3:

The eccentricity of the rotor generates a dynamical force:

   $F(t) = m_1 e\Omega^2\cos\Omega t$ with $\Omega = 157.1\,\text{s}^{-1}$

The force transmitted to the foundation is $Q(t) = cq(t)$

   $V(\eta) = \dfrac{Q_{max}}{F_{max}} = 0.2 \quad \rightarrow \quad \eta = \sqrt{6} \quad \rightarrow \quad \omega_0 = \sqrt{\dfrac{c}{m}} = 64.14\,\text{s}^{-1}$

   $c \le 4.113\,\text{kN/mm}$, $\hat{q} = 0.036\,\text{mm}$, $q_{st} = 2.39\,\text{mm}$

Problem 11.4:

   $V_a = \dfrac{1}{1 + \eta^2}$, $V_b = 2D\eta V_a$

Chapter 12: Systems of Several Degrees of Freedom

Problem 12.1:

   $\omega_1 = 0.807\sqrt{\dfrac{EJ}{ml^3}}$, $\omega_2 = 2.82\sqrt{\dfrac{EJ}{ml^3}}$

Problem 12.2:

$$\omega_1^2 = \frac{g}{l}, \quad \omega_2^2 = \frac{g}{l} + \frac{2ca^2}{ml^2}$$

Problem 12.3:

1. $\dfrac{5}{12}\ddot{q}_1 + 4\lambda q_1 + \lambda q_2 = 0$

    $\dfrac{5}{2}\ddot{q}_2 + 6\lambda q_1 + 4\lambda q_2 = 0, \quad \lambda = \dfrac{EJ}{ml^3}$

2. $\left(4\lambda - \dfrac{5}{12}\omega^2\right)q_1 + \lambda q_2 = 0$

    $6\lambda q_1 + \left(4\lambda - \dfrac{5}{2}\omega^2\right)q_2 = 0$

3. $\omega_{1,2}^2 = \lambda\left\{\dfrac{1}{5}\left(28 \pm \sqrt{544}\right)\right\}$

Problem 12.4:

See 12.3.

Problem 12.5:
The system is of 2 DOF. Following Eq. (11.19), we find

$$q_1 = \delta_{11}X_1 + \delta_{12}X_2 = -m\ddot{q}_1\delta_{11} - \theta\ddot{q}_2\delta_{12}$$
$$q_2 = \delta_{21}X_1 + \delta_{22}X_2 = -m\ddot{q}_1\delta_{21} - \theta\ddot{q}_2\delta_{22}$$

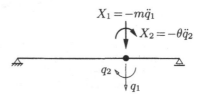

with

$$EJ\delta_{11} = \frac{4}{9}l^3, \quad EJ\delta_{12} = EJ\delta_{21} = -\frac{2}{9}l^2, \quad EJ\delta_{22} = \frac{1}{3}l.$$

The mass moment of inertia is

$$\theta \approx \frac{1}{16}ml^2.$$

Thus, from the Lagrangian determinant (Eq. 12.10)

$$\omega_{1,2} = \begin{cases} 33,27\ \mathrm{s}^{-1} \\ 191,2\ \mathrm{s}^{-1} \end{cases}$$

The modes are determined from the ratio $\hat{q}_2/\hat{q}_1$

$$\left(\frac{\hat{q}_2}{\hat{q}_1}\right)_{1,2} = \frac{1 - \delta_{11}m\omega^2}{\delta_{12}\theta\omega^2} = \begin{cases} -0,514 \\ 31,02 \end{cases}$$

Problem 12.6:

$$E = \frac{1}{2}\dot{q}_1^2(m_1 + m_2 + m_3) + \dot{q}_1\dot{q}_2 l m_2 \cos q_2 + \dot{q}_1\dot{q}_3 m_3$$

$$+ \frac{1}{2}\dot{q}_2^2 l^2 m_2 + \frac{1}{2}\dot{q}_3^2 m_3$$

$$\Pi = m_2 g l(1 - \cos q_2) + \frac{1}{2}c q_3^2$$

1. $L = E - \Pi$
2. $H = E + \Pi$

3. $\ddot{q}_1 = a, \quad \ddot{q}_2 + \frac{1}{l}(g \sin q_2 + a \cos q_2) = 0, \quad \ddot{q}_3 + \frac{c}{m_3} q_3 = -a$

Index